SpringerBriefs in Research and Innovation Governance

Editor-in-Chief

Doris Schroeder, Centre for Professional Ethics, University of Central Lancashire, Preston, Lancashire, UK

Konstantinos Iatridis, School of Management, University of Bath, Bath, UK

SpringerBriefs in Research and Innovation Governance present concise summaries of cutting-edge research and practical applications across a wide spectrum of governance activities that are shaped and informed by, and in turn impact research and innovation, with fast turnaround time to publication. Featuring compact volumes of 50 to 125 pages, the series covers a range of content from professional to academic. Monographs of new material are considered for the SpringerBriefs in Research and Innovation Governance series. Typical topics might include: a timely report of state-of-the-art analytical techniques, a bridge between new research results, as published in journal articles and a contextual literature review, a snapshot of a hot or emerging topic, an in-depth case study or technical example, a presentation of core concepts that students and practitioners must understand in order to make independent contributions, best practices or protocols to be followed, a series of short case studies/debates highlighting a specific angle. SpringerBriefs in Research and Innovation Governance allow authors to present their ideas and readers to absorb them with minimal time investment. Both solicited and unsolicited manuscripts are considered for publication.

More information about this series at http://www.springer.com/series/13811

Katharina Jarmai

Editor

Responsible Innovation

Business Opportunities and Strategies
for Implementation

Editor
Katharina Jarmai
Institute for Managing Sustainability
WU Vienna University of Economics and Business
Vienna, Austria

ISSN 2452-0519   ISSN 2452-0527   (electronic)
SpringerBriefs in Research and Innovation Governance
ISBN 978-94-024-1719-7   ISBN 978-94-024-1720-3   (eBook)
https://doi.org/10.1007/978-94-024-1720-3

This Springer imprint is published by the registered company Springer Nature B.V.
The registered company address is: Van Godewijckstraat 30, 3311 GX Dordrecht, The Netherlands

Acknowledgments

The work that this book is based on was the collaborative effort of all the consortium members of the Horizon 2020 project "COMPASS – Evidence and Opportunities for Responsible Innovation in SMEs". I believe I can speak for everybody involved by saying that the work in this project was immensely satisfying due to the fact that all members of the project consortium were highly professional and, at the same time, tremendously enjoyable to work with. Therefore, I would like to express my gratitude both to the authors of this book and to those members of the consortium that were busy undertaking other COMPASS project work in the meantime (in alphabetical order of partner organization name): Nathan Gilbert at B Lab Europe, Chiara Davalli at the European Business and Innovation Centre Network, Alexandre Estéban at "la Caixa" Banking Foundation, and Marcelline Bonneau, Christophe Gouache, and François Jégou at Strategic Design Scenarios.

I would like to thank all the authors for investing time and effort into this project, keeping deadlines, remaining open to suggestions, and providing constructive feedback to their fellow authors.

Thank you to Julie Cook, who has provided us not only with her top-notch professional proofreading services but also with her knowledgeable suggestions on how to improve the comprehensibility of each chapter.

I would further like to thank Fritz Schmuhl at Springer for organizing the smooth publishing process and to the two anonymous reviewers for providing useful comments on abstracts and final drafts.

Last but not least, I am grateful to Prof. Doris Schröder for initiating the project of writing this book and for gently guiding me through the production process.

Contents

Contributors

Josephina Antoniou University of Central Lancashire, School of Sciences, Pyla, Cyprus

Catherine Flick School of Computer Science and Informatics, De Montfort University, Leicester, UK

Malcolm Fisk School of Computer Science and Informatics, De Montfort University, Leicester, UK

Katharina Jarmai Institute for Managing Sustainability, WU Vienna University of Economics and Business, Vienna, Austria

André Martinuzzi Institute for Managing Sustainability, WU Vienna University of Economics and Business, Vienna, Austria

Caroline Nwafor Institute for Managing Sustainability, WU Vienna University of Economics and Business, Vienna, Austria

George Ogoh School of Computer Science and Informatics, De Montfort University, Leicester, UK

Norma Schönherr Institute for Managing Sustainability, WU Vienna University of Economics and Business, Vienna, Austria

Doris Schroeder School of Law, University of Central Lancashire, Pyla, Cyprus

Adele Tharani Institute for Managing Sustainability, WU Vienna University of Economics and Business, Vienna, Austria

Abbreviations

CEO   Chief Executive Officer
CS   Corporate Sustainability
CSO   Civil Society Organization
CSR   Corporate Social Responsibility
DK   Denmark
EC   European Commission
EE   Estonia
EMAS   Eco-Management and Audit Scheme
EPSRC   Engineering and Physical Sciences Research Council (UK)
ESG   Environmental, Social, and Governance Issues
FI   Finland
GDPR   General Data Protection Regulation
GMO   Genetically Modified Organism
ICN2   Catalan Institute of Nanoscience and Nanotechnology
ICREA   Institut Català de Recerca i Estudis Avançats
ICT   Information and Communication Technologies
ISO   International Organization for Standardization
LV   Latvia
NGO   Non-governmental Organization
NL   Netherlands
R&D   Research and Development
RI   Responsible Innovation
RRI   Responsible Research and Innovation
SE   Sweden
SOI   Sustainability-Oriented Innovation
SME   Small- and Medium-Sized Enterprises
UAB   Universitat Autònoma de Barcelona
UK   United Kingdom

Chapter 1
Introduction

Katharina Jarmai

Abstract The concept of Responsible Research and Innovation (RRI) originates in discourses on emerging technologies and research ethics in contested innovative fields, such as nanotechnologies or geo-engineering, and has been predominantly driven by European research and innovation policy over the past 10 years. The concept was initially developed and introduced by policy makers and social scientists, but recent studies have aimed to shed light on the implementation of responsible research and innovation practices in business. The contributions collected in this book are a result of work conducted by seven partner organisations in the European funded Horizon 2020 project "COMPASS – Evidence and opportunities for responsible innovation in SMEs". In combination, they illustrate that responsible innovation (RI) has been emerging as a new field in the ongoing discourse on the role and responsibility of business in society.

Keywords Responsible innovation · RRI · RRI introduction · Responsible business · COMPASS project

1.1 A Brief Introduction to Responsible (Research and) Innovation

The concept of Responsible Research and Innovation (RRI) originates in discourses on emerging technologies and research ethics in contested innovative fields, such as nanotechnologies or geo-engineering, and has been predominantly driven by European research and innovation policy since 2011 (Owen et al. 2012). A first working definition of RRI was proposed by von Schomberg (2011: 9) as:

> "[a] *transparent, interactive process by which societal actors and innovators become mutually responsive to each other with a view on the (ethical) acceptability, sustainability and*

K. Jarmai (✉)
Institute for Managing Sustainability, WU Vienna University of Economics and Business, Vienna, Austria
e-mail: katharina.jarmai@wu.ac.at

© The Author(s) 2020
K. Jarmai (ed.), *Responsible Innovation*, SpringerBriefs in Research and Innovation Governance, https://doi.org/10.1007/978-94-024-1720-3_1

societal desirability of the innovation process and its marketable products (in order to allow a proper embedding of scientific and technological advances in our society)".

After a period of debate about the definition of RRI (Stilgoe et al. 2013) and the concept's continuous development (Blok and Lemmens 2015), a common, general agreement about the meaning and key aspects of RRI has developed in the form of the four dimensions of anticipation, reflection, inclusion/deliberation and responsiveness (Stahl et al. 2017). Integration of these four dimensions in research and innovation processes should lead towards more responsible innovation output (Owen et al. 2012, 2013; Stilgoe et al. 2013). At the same time, the European Commission has been promoting RRI by funding projects on the thematic elements of ethics, gender and diversity, public engagement, open access, and science education through the previous and current European Framework Programmes for Research and Innovation, "FP7" and "Horizon 2020". For the upcoming Framework Programme "Horizon Europe" (2021–2027), the European Commission proposes that the programme "[...] should engage and involve citizens and civil society organisations in co-designing and co-creating responsible research and innovation agendas and content, promoting science education, making scientific knowledge publicly accessible, and facilitating participation by citizens and civil society organisations in its activities."; both across the programme and through dedicated activities (European Commission 2018: para 26). It remains to be seen to what extent the different elements of the RRI concept as described above, and its basic aim to increase positive societal impact and minimize potential risks for individuals, the society and the natural environment will be implemented in the Horizon Europe programme and in the research and innovation projects it funds.

The RRI concept was initially developed and introduced by policy makers and social scientists (Lubberink et al. 2017), but recent studies have aimed to shed light on the implementation of RRI practices in business. These studies indicate that businesses in Europe still seem to be operating without an awareness of the concept itself (Blok and Lemmens 2015; Davies and Horst 2015; Khan et al. 2016), but that extant practices, processes and purposes exhibit indications of responsible innovation (Asante et al. 2014). Moreover, a growing body of literature has been dealing with questions of how to incentivise or drive companies to adopt either the concept (Auer and Jarmai 2018; Gurzawska et al. 2017; Chatfield et al. 2017), or particular responsible innovation principles (Iatridis and Kesidou 2018; Iatridis and Schroeder 2016). First good practice examples of implementation of RRI in business provide a diverse set of company practices; ranging from inclusive governance and a general orientation of company research and innovation towards tackling societal challenges, through institutionalized opportunities for anticipation and reflection, to targeted activities aimed at increasing gender balance or fostering science education (Schroeder 2014, 2017).

The discourse about embedding responsibility in corporate innovation processes has evolved from aiming to implement the RRI concept, as defined by European Commission, to linking it to extant responsibility concepts such as Corporate Social Responsibility (CSR) or tangible company practices throughout the innovation process. In the course of these developments, the use of the simpler term "responsible innovation" (RI) has emerged, which has been used synonymously with the abbreviation "RRI". The term responsible innovation is more common in communities that deal with responsibility in corporate innovation processes but are not directly influenced by the European Commission's Research Programmes. This is reflected, for example, by the launch of a Special Interest Group for Responsible Innovation by the International Society for Professional Innovation Management (ISPIM)[1], the foundation of the "Virtual Institute for Responsible Innovation"[2] hosted by the Center for Nanotechnology in Society at Arizona State University in the US and, last but not least, the title of the first academic journal concentrating on the assessment and governance of innovation, namely the Journal of Responsible Innovation[3].

1.2 Business Opportunities Through Responsible Innovation? A Response in Six Chapters

The contributions collected in this book are the results of work conducted by seven partner organisations in the European funded Horizon 2020 project "COMPASS"[4]. The overall objective of the project was to develop tools to support Small and Medium-Sized Enterprises (SMEs) in the implementation of RI. Main project outputs include an online self-check tool that allows companies to find out what they already do that qualifies as RI and what other actions they can take, a methodology to develop a company-specific action plan for RI, and sector-tailored roadmaps for companies working with nanotechnologies, in cyber security, or in biomedicine. In the process of co-developing these tools together with companies, sector experts, funding organisations and civil society representatives, the members of the project consortium used their expertise to support companies that were looking for ways to increase their positive impact on society while at the same time aiming to discover the benefits of applying RI principles that are measurable in terms of revenue.

This book is tailored towards the interests of innovation managers, entrepreneurs and academics. For innovation managers and entrepreneurs, it will provide inspiration and ideas about what RI can look like in practice and what the possible benefits might be. For readers with an academic interest, it offers discussion of potential company incentives for RI and suggestions as to how to communicate its essence to companies in a useful and comprehensive way. In Chap. 2, Jarmai and colleagues tackle the challenge of embedding the responsible innovation concept in a business context and suggest a five step strategy on how a company can engage with RI. In Chaps. 3 and 4, Jarmai and Antoniou, respectively, connect RI to sustainability-

---

[1] https://www.ispim-innovation.com/responsible-innovation
[2] https://cns.asu.edu/viri
[3] https://www.tandfonline.com/action/journalInformation?show=aimsScope&journalCode=tjri20
[4] https://innovation-compass.eu/

oriented innovation and social innovation to illustrate what responsible innovation can learn from connecting with these two approaches, which are better established in a business context. In Chap. 5, Schroeder presents real-life case studies of companies that have implemented RI practices, and discusses reported benefits. In Chap. 6, Flick and colleagues give a first-hand account of challenges and successful strategies for co-creating RI strategies with companies. In the concluding Chap. 7, Schönherr and colleagues summarise the most important lessons learned from the contributions in this volume, and develop the outlines of a business case for responsible innovation.

In combination, the chapters in this volume illustrate that responsible innovation is emerging as a new field in the continuing discourse on the role and responsibility of business in society. Success in economic terms cannot be guaranteed, but the willingness of an SME to innovate in areas that have positive societal impact in addition to profits can bring business benefits and add additional value such as higher employee satisfaction, retention of skilled personnel or reputational gains.

References

Asante, K., Owen, R., & Williamson, G. (2014). Governance of new product development and perceptions of responsible innovation in the financial sector: insights from an ethnographic case study. *Journal of Responsible Innovation, 1*(1), 9–30.

Auer, A., & Jarmai, K. (2018). Implementing responsible research and innovation practices in SMEs: Insights into drivers and barriers from the Austrian medical device sector. *Sustainability, 10*(1), 17.

Blok, V., & Lemmens, P. (2015). The emerging concept of responsible innovation. Three reasons why it is questionable and calls for a radical transformation of the concept of innovation. In B.-J. Koops, I. Oosterlaken, H. Romijn, T. Swierstra, & J. van den Hoven (Eds.), *Responsible innovation 2: Concepts, approaches, and applications* (pp. 19–35). Cham: Springer.

Chatfield, K., Iatridis, K., Stahl, B. C., & Paspallis, N. (2017). Innovating responsibly in ICT for ageing: Drivers, obstacles and implementation. *Sustainability, 9*(6), 971.

Davies, S. R., & Horst, M. (2015). Responsible innovation in the US, UK and Denmark: Governance landscapes. In B.-J. Koops, I. Oosterlaken, H. Romijn, T. Swierstra, & J. van den Hoven (Eds.), *Responsible innovation 2: Concepts, approaches, and applications* (pp. 37–56). Cham: Springer.

European Commission. (2018). Proposal for a REGULATION OF THE EUROPEAN PARLIAMENT AND OF THE COUNCIL establishing Horizon Europe – the Framework Programme for Research and Innovation, laying down its rules for participation and dissemination. COM(2018) 435 final. https://eur-lex.europa.eu/resource.html?uri=cellar:b8518ec6-6a2f-11e8-9483-01aa75ed71a1.0001.03/DOC_1&format=PDF. Accessed 18 Dec 2018.

Gurzawska, A., Mäkinen, M., & Brey, P. (2017). Implementation of responsible research and innovation (RRI) practices in Industry: Providing the right incentives. *Sustainability, 9*, 1759.

Iatridis, K., & Kesidou, E. (2018). What drives substantive versus symbolic implementation of ISO 14001 in a time of economic crisis? Insights from Greek manufacturing companies. *Journal of Business Ethics, 148*(4), 859–877.

Iatridis, K., & Schroeder, D. (2016). *Responsible research and innovation in Industry. The case for corporate responsibility tools*. Heidelberg: Springer.

Khan, S. S., Timotijevic, L., Newton, R., Coutinho, D., Llerena, J. L., Ortega, S., Benighaus, L., Hofmaier, C., Xhaferri, Z., de Boer, A., Urban, C., Strähle, M., Da Pos, L., Neresini, F., Raats, M. M., & Hadwiger, K. (2016). The framing of innovation among European research funding actors: Assessing the potential for 'responsible research and innovation' in the food and health domain. *Food Policy, 62*, 78–87.

Lubberink, R., Blok, V., van Ophem, J., & Omta, O. (2017). Lessons for responsible innovation in the business context: A systematic literature review of responsible, social and sustainable innovation practices. *Sustainability, 9*(5), 721.

Owen, R., Macnaghten, P., & Stilgoe, J. (2012). Responsible research and innovation: From science in society to science for society, with society. *Science and Public Policy, 39*(6), 751–760.

Owen, R., Stilgoe, J., Macnaghten, P., Gorman, M., Fisher, E., & Guston, D. (2013). A framework for responsible innovation. In R. Owen, J. Bessant, & M. Heintz (Eds.), *Responsible innovation, managing the responsible emergence of science and innovation in society* (pp. 51–74). Chichester: Wiley.

Schroeder, D. (2014). *D1.2 Case study descriptions. Deliverable of the FP7 project RESPONSIBLE-INDUSTRY.* http://docs.google.com/viewer?a=v&pid=sites&srcid=ZGVmYXVsdGRvbW-FpbnxyZXNwb25zaWJsZWluZHVzdHJ5d2Vic2l0ZXxneDoyZjdkYmZkNWJmMzVhY-zkx

Schroeder, D. (2017). *D1.2 case study descriptions.* Deliverable of the Horizon 2020 project COMPASS. https://innovation-compass.eu/wp-content/uploads/2017/07/Deliverable-1_3-Compass-Case-Study-Descriptions.pdf. Accessed 20 Sept 2018.

Stahl, B., Obach, M., Yaghmaei, E., Ikonen, V., Chatfield, K., & Brem, A. (2017). The responsible research and innovation (RRI) maturity model: Linking theory and practice. *Sustainability, 9*(6), 1036.

Stilgoe, J., Owen, R., & Macnaghten, P. (2013). Developing a framework for responsible innovation. *Research Policy, 42*(9), 1568–1580.

von Schomberg, R. (2011). *Towards responsible research and innovation in the information and communication technologies and security technologies fields.* Directorate general for research and innovation. https://doi.org/10.2139/ssrn.2436399.

Chapter 2
Responsible Innovation in Business

Katharina Jarmai, Adele Tharani, and Caroline Nwafor

Abstract This chapter introduces responsible innovation in a business context. The first part explains the basic terms that constitute responsible innovation from a business perspective. The second part presents tangible business practices that operationalise responsible innovation and introduces two good practice examples that hint at the variety of ways in which responsible innovation can be implemented in companies.

Keywords Responsible research and innovation · Responsible innovation · Corporate social responsibility · Applied nanoparticles · Yoti · B Corporation

2.1 Introduction

"So, do you mean that I am irresponsible?"

This is the response you may get when you ask an entrepreneur if they would like to make their company's innovation processes and innovative products more responsible. Once you start explaining the elements of the responsible innovation (RI) concept, your conversation partner will likely relax and confirm that yes, consumer trust, ethical conduct or safety considerations are indeed of interest to their company, and that yes, they would be interested in hearing more about how they can decrease the risk of failing to meet consumer wants, or being blamed for undesirable side-effects of her company's innovation at a later point in time.

This chapter presents the contents of the conversation that could follow. To break down the concept of RI into practices that make sense in a business context, we first explore the two elements of RI, i.e. responsibility and innovation, from a business management perspective (Sect. 2.2). We then present RI as a collection of tangible

K. Jarmai (✉) · A. Tharani · C. Nwafor
Institute for Managing Sustainability, WU Vienna University of Economics and Business, Vienna, Austria
e-mail: katharina.jarmai@wu.ac.at; adele.tharani@wu.ac.at; caroline.nwafor@wu.ac.at

© The Author(s) 2020
K. Jarmai (ed.), *Responsible Innovation*, SpringerBriefs in Research and Innovation Governance, https://doi.org/10.1007/978-94-024-1720-3_2

company practices (Sect. 2.3) and introduce two companies that have already implemented many of these practices in their own particular way (Sects. 2.3.1 and 2.3.2). To conclude, we summarize main learning about RI in a business context (Sect. 2.4).

2.2 Defining "Responsibility" and "Innovation" in a Business Context

2.2.1 Responsibility

Business responsibility towards society has a longer history in business management literature than the idea of responsible innovation, or responsibility of science towards society. For a long time primary responsibility of business was defined only in economic terms – responsibility towards shareholders and the responsibility to make profit. The discourse on the extension of business' responsibility to stakeholders and broader society can be traced back to the 1950s and 1960s (Carroll and Shabana 2010), with scholars such as Howard R. Bowen (1953) and Peter Drucker (1954), who discussed the moral and ethical responsibilities of a business, and as such a business manager, towards society and the public good. The responsibility of business towards society has carried a number of conceptualisations, including philanthropy, business ethics, corporate social responsibility, corporate citizenship and corporate sustainability (Carroll and Shabana 2010). Corporate social responsibility (CSR) and corporate sustainability (CS) are currently the more common terms in business practice and are showing signs of convergence (Montiel 2008), yet so far have no fixed standardised definition (Montiel and Delgado-Ceballos 2014). In essence, business' responsibility to society can be linked to three main theories: stakeholder theory, social contracts theory and legitimacy theory (Moir 2001).

The concept of business responsibility has evolved from the philanthropic approach of "giving back", to a more strategic approach to business' responsibility towards society being addressed in management literature. Since the early 2000s scholars have started to connect business' strategic economic goals with business' roles and responsibilities towards society, with numerous studies examining the "business case of CSR" (Carroll and Shabana 2010). Porter and Kramer (2011) argue that by acting responsibly and gearing a business towards responding to societal needs, business can simultaneously serve its economic and societal responsibility and introduced the idea that business is a force that can "create shared value". Their idea brought business responsibility from the fringes of the company to the core of business strategy (Crane et al. 2014). Company responsibility was no longer seen as an activity outside a company's core operations and core competencies, but rather as responsiveness to societal needs through creating products and services, which became a potential avenue for business growth. One can argue that this shift from responsibility as an afterthought to responsibility as a strategy also fuelled the increasing replacement of the term corporate social responsibility with corporate sustainability. The latter signifies that responding to societal needs and acting responsibly towards people and the environment is a precondition to business survival.

Recent years have seen a strong societal push to acknowledge that businesses' value chains, from sourcing of raw materials to production, sales and product end-of-life, cause impacts on people and the environment for which they are responsible. Therefore, the European Commission, as well as other public actors, has redefined what CSR means, from a "company voluntary contribution to society", to company "responsibility for its impacts" on society (European Commission 2011), including people and the environment.

With these societal pressures, the understanding and conceptualisation of company responsibility towards society now encompasses a number of issues, themes and business activities. Companies from a variety of sectors are being scrutinised for their effects on people and the environment throughout their value chains (Phillips and Caldwell 2005); extending their responsibility for impact beyond their own operations to supply chains and product use. This includes materials sourcing and procurement in supply chains, production, transport and packaging, and lastly the life-cycle effects from the actual use of the product, and its disposal or afterlife. The issues range from environmental resource use or emissions into air, land or water, to effects on human rights, ensuring decent work and health and safety in company production or operations, ensuring an environmentally friendly afterlife of company products or even the social desirability of a company's products and services. Therefore, business responsibility towards society now means both, responsible management of business operations, as well as a business' responsibility for the impacts of its products and services on people and the environment. Companies globally are being expected to take responsibility for doing no harm to people or the environment, whereas the most advanced ones are looking into strategies that drive the business through responsiveness to societal needs.

2.2.2 Innovation

The story of innovation often begins with the economist Joseph Schumpeter (1883–1950) and is thus deeply rooted in socio-economic theory. For Schumpeter (1939), economic development was a dynamic process driven by the development of novel[1] combinations – innovations – which in processes of creative destruction

---

[1] Definitions of innovation differentiate between the scopes of novelty. While Kieser (1969) defines innovation as novelty at the level of an organization, according to Vedin (1980: 22) innovation is "…an invention brought to its first use, its first introduction to the market". Garcia and Calantone (2002) identify six perspectives of novelty in the innovation literature current at the time: New to the company, new to the adopting unit, new to the market, new to the industry, new to the consumer and new to the world.

generate new business models. Innovation can arise in the form of a new product formerly unknown to the consumer, but also in the form of a new quality of an existing product. Likewise, innovation can emerge in the form of the introduction of a new production method, an opening up of new sales markets, the development of new sources of raw materials or a re-organization of a business already in the market. In any form, innovation allows businesses to occupy a temporary monopoly position, which lasts until competing businesses either successfully imitate the innovation or gain supremacy through further or novel developments.

From an economic point of view, innovation is generally conceived as the basis for a competitive economy (cf. Adams et al. 2006) and thus as something that is inherently desirable in the present perception of the western industrialized world (cf. Blok and Lemmens 2015; Moldaschl 2010). Companies pursue innovation to develop new market segments, improve the quality of their products or reduce the costs of production. They aim to maintain their competitive edge or improve their position in the market through innovative products (goods and services), innovative processes (production or delivery methods), innovative marketing (design, packaging, placement, promotion, pricing) or organisational innovation (business practices, workplace organisation, external relations). In this constant race for novelty and improvement only those that constantly reinvent themselves and their products can win. An innovation's success is, however, measured in terms of its uptake on the market and its generation of economic profit for the owner of the innovation. Societal benefit may arise as positive externalities of innovation, but are not per-se decisive for action. In this way, innovations can be a source of income for the innovation owner and at the same time lead to job losses, or cause short- or long-term environmental, health or safety issues that may or may not become apparent at the time of the innovation's introduction to the market. This fact has found its countermovement in approaches to substitute solutions on the market with more eco-friendly, more sustainable or more socially desirable ones (see Chaps. 3 and 4 for an introduction to these types of innovation), and thus combine the pursuit of competitiveness with a normative requirement to reduce harm to people and the natural environment.

Innovation management in companies is mostly concerned with creating fruitful environments for new ideas, and deciding which of these ideas will be pursued further and which are to be discarded. This means that not every new idea will necessarily turn into, or lead to, innovation[2]. It also means that innovation management is constantly concerned with creating opportunities for innovation through the formulation of new ideas, and destroying opportunities for innovation by discarding a large proportion of these ideas before they reach the market, or even the develop-

---

[2] Different authors consider different events as decisive for defining innovation: According to Roberts (1987), innovation takes place when an introduction to the market is followed by commercial exploitation, application, diffusion and further development; an innovation needs to be successful on the market and create value in order to earn the name. Brockhoff (1992), in contrast, considers market entry to be sufficient criterion for product or process innovation; irrespective of its level of commercial success.

ment phase. Management literature has extensively discussed approaches to establishing an innovation culture within an organization, the pros and cons of involving company external actors and the selection process of one idea over another. These discussions are too versatile to be reproduced in this chapter. It should be noted, however, that decisions to pursue one idea and discard another are generally taken under high levels of uncertainty about potential success. The higher the dynamics in a particular market and the more radical[3] the innovation, the higher the levels of uncertainty will be. Well-known management approaches to decrease uncertainty through utilising information from company-external sources include open innovation, user innovation and innovation communities (Fichter 2009; Chesbrough et al. 2006; Gassmann and Enkel 2004; von Hippel 1986).

In contrast to a few decades ago, today a company can fall back on various methods to support both the idea generation process as well as the selection of ideas for further development. Many of these methods have been developed by companies and now find their way into both the practitioner and academic literatures. These range from "classic" methods such as brainstorming to more recent development such as design thinking, the "innovation sprint" (cf. e.g. Ma and Morris 2017), or gaming approaches.

2.3   How Should Companies Be(come) Responsible in Their Innovation Activities?

Once an entrepreneur's interest in responsible innovation has been stirred, they will probably have two pressing questions: 'What exactly do I need to do?' and 'What's in it for me and my business?' Innovation is usually closely connected to the core business of a company, and different companies operate under different conditions (depending on e.g. business model, size, product and contextual factors such as legal frameworks or sector dynamics), so there are no universally valid answers to these questions. It is, however, possible to describe a general process to develop a tailored RI strategy and point out resources that companies can rely on. The following five steps provide guidance to a company wishing to engage with RI:

1. **Understand what responsible innovation is all about.** While you may not have heard of the concept of responsible innovation, your company may already be doing things that fall under the concept of RI. You can figure out in which areas of your operations RI might be particularly important, find out what your current strengths are, and what action you may want to take. One way to self-assess your

---

[3] While incremental innovations are born along existing paths of (technological) development and improve the performance of existing products or processes, radical innovations are disruptive in their nature and create path changes. Radical innovation is generally followed by a multitude of incremental innovations, and often by organisational or societal changes (cf. e.g. Utterback 1996).

company is using the COMPASS online self-check tool[4]. Another way is reflecting together with a responsible innovation expert.

2. **Reflect on the expected benefits of responsible innovation.** The implementation of RI will take time, may require additional investments and will likely require changes in company practices. It could result in the (re-)definition of company values, goals or collaboration patterns. It might even affect the company's business model, if people realize that core business activities are not in line with the objectives of RI. A company will only invest in these efforts if a particular added value can be expected. This added value can be measured in terms of e.g. improving customer relationships, pro-actively meeting expected future regulation, or increasing the company's positive impact on society. This will differ between sectors, regions and individual companies. The crucial point is to understand what pursuing RI may yield and what the company is willing to invest to this end.

3. **Establish management and employee commitment.** To ensure that time and money is allocated to employees' engagement with RI, and that practices are actually implemented, both top management as well as employees need to commit to pursuing RI. Such commitment can, for example, be facilitated through inclusive development of a company Code of Conduct that respects RI (see Sect. 2.3.2), or providing employees with training in RI.

4. **Develop an action plan for development/adaptation of practices**. Once a company has a clear idea of where it stands, and commits to making a step towards RI, an action plan can be developed. First, identify contextual factors that will likely shape your company's working context, potential markets, societal trends, workforce and collaborations in the middle- to longer-term future. Then identify practices and milestones, and specify responsibilities and deadlines. Potentially, you could develop indicators and procedures for monitoring progress. If you already apply suitable methods for taking these steps in your company, make use of those. If you are not familiar with any suitable methods, you could utilize e.g. the COMPASS co-creation method kit[5] or procure the services of a consultant or facilitator.

5. **Stay focused on the objective of responsible innovation**. Different aspects of RI will seem more relevant than others, depending on the company and the context it operates in. Some practices will be more intriguing in terms of expected added value. Nevertheless, it is important to implement practices that are the most important and critical in that sector, and which cover different aspects of RI

---

[4]The COMPASS self-check tool allows companies to find out what they already do that qualifies as responsible innovation and what they could do to improve their responsible innovation performance. All proposed practices are entirely within company control and can be put into practice one-by-one or in combination. The tool has been available free of charge at https://innovation-compass.eu/self-check/ since March 2019.

[5]The COMPASS co-creation method kit provides detailed instructions on how to conduct (a) a forward-looking exercise to identify future relevant company context and important responsible innovation practices and (b) a back casting exercise to develop an actionable roadmap for the company. It has been publicly available at https://innovation-compass.eu/method-kit/ since March 2019.

to keep pursuing the overall objective, which is to increase positive societal impact and minimize actual and potential negative impact to the highest degree possible. Keep re-evaluating your company practices and adapting your action plan at regular intervals to respond to changing contextual factors, technological advances and company developments.

The next two sections of this chapter present two companies that have success-fully completed these five steps towards responsible innovation[6]. The two good practice examples hint at the variety of ways in which RI can be implemented in companies. One of them details the various practices that a nanotechnology com-pany has introduced to ensure that all of its research and innovation processes and products exceed the requirements of RI, the other demonstrates how a cyber security company relies on the principles of RI to inform their decision-making processes.

2.3.1 Good Practice Example 1: Responsible Innovation as a Business Model[7]

Applied Nanoparticles SL (AppNps) was founded in 2013, arising as a spin-off company from the Universitat Autònoma de Barcelona (UAB), the Institut Català de Recerca i Estudis Avançats (ICREA), and the Catalan Institute of Nanoscience and Nanotechnology (ICN2), with the goal to base the research and development of nanoparticles on Responsible Research and Innovation (RRI) principles. AppNps's main product is BioGAS+, which is an additive based on iron nanoparticles directed to the optimisation of anaerobic digestion processes. The main aim of BioGAS+ is to transform waste into appealing raw materials in an efficient and sustainable way.

AppNps's company structure exhibits several features that demonstrate its com-mitment to the principles of responsible innovation. There is a collective ownership of the company without an explicit CEO. Employees are involved in decision-making with the objective of keeping the company diverse and robust; and to ensure that the initial aims of the company are preserved. This is what AppNps refers to with their slogan "a company in the making". Their second slogan, "a company with purpose", refers to the collectively agreed upon vision to become a role model in terms of responsible innovation and nanoparticles. Aware of the need for communication between science and society for a smooth introduction of nanotechnology in society, AppNps is a strong advocate of science education and a pronounced stakeholder

---

[6] Both cases were developed according to the requirements of Sage Business Cases (http://sk.sage-pub.com/cases) and will be published in 2020. Each case consists of an introduction to nanotech-nology or cyber security, respectively, an introduction to responsible innovation, the case, expected learning outcomes and discussion questions. Both cases are further accompanied by a teaching note that describes teaching objectives, target audience, suggested teaching strategy and suggested answers to the discussion questions.

[7] The complete case study is available at https://innovation-compass.eu/training/cases/

dialogue. Efforts to act responsibly are further implemented through implementation of safety and health regulations that go beyond compliance and employee's education on nanotoxicity and nanosafety. On the whole, AppNps continually monitors that actions are directed to seek social, economic and environmental sustainability.

AppNps constantly need to balance scientific and commercial interests and tackle the challenge of involving society in a way that benefits both society and the company. The company is willing to engage themselves with ethical dilemmas, such as the ethical dimension of growing crops when this could potentially affect food security negatively. These questions require AppNps to constantly reflect and re-evaluate its values and strategies.

2.3.2   Good Practice Example 2: Responsible Innovation as a Decision Support System[8]

Yoti[9] is a London-based information technology company which was founded in 2014. It employed over 200 people in 2018 and has offices in India, the US and Canada. Yoti brings together the advance in biometric technologies and an increased smartphone usage to create a digital identity solution that allows online users to prove who they are without compromising their privacy. More specifically, the main company product in an app that combines biometric information with a government-issued identity document (passport, driving license, etc.). The app was launched in 2017 and was downloaded more than 1.5 million times within the first half year. The Yoti app is beneficial both for organisations – to verify online and in person who people are – and for individuals – to prove their age or identity with their smartphone.

Yoti aims to have a positive impact on society and has the goal to become the world's most trusted identity platform. The firm is well aware of the responsibility that goes along with handling personal data and thus considers data responsibility to be a core strategy of the business model. This is achieved by asking users to provide only a minimal amount of data in the first place and by implementing a system that encrypts and stores the data separately, so that only the individual users can tie together all the data. Moreover, Yoti has put several principles in place to ensure a maximum amount of transparency and consumer trust. These principles include continually considering the firm's impact on users, employees, suppliers, partners and the environment; providing a digital identity to anyone for free; and disclosing terms and conditions in a transparent way. To watch over the compliance with these

---

[8] As a cyber security B Corporation based in the UK, Yoti was invited to participate in the COMPASS project in 2017. Yoti was highly active in co-creating the Responsible Innovation roadmap for cyber security, and has since been in close contact with the COMPASS Consortium for advice on questions related to ethics and transparency, and to spread word about data trust and security issues.

[9] https://www.yoti.com/

principles, Yoti has installed a "Guardian Council"[10]. The principles are practically applied by an intense stakeholder collaboration, an engagement with digital policy, digital identity and data protection advocates and by committing to standards beyond legal requirements.

Inherent to the collection and storage of data there are multiple challenges. First and foremost, this is the issue of data privacy. In contrast, from a perfect privacy of user data arises the possible implication of facilitating illegal activity. Minimizing the risk of criminal misuse while maintaining data security is a major challenge that has been tackled in an extensive dialogue with human rights and consumer rights experts.

2.4 Chapter Conclusions

Responsible innovation, albeit born in the public policy realm and to date adopted primarily by research institutions, is highly complementary to the broad concept of business responsibility towards society. From the process dimension it allows the extension of the concept of responsible business management to research and development (R&D) departments, and guides businesses in how to make their R&D more responsible and responsive to societal needs. Traditional approaches to CSR have not as yet extensively considered the R&D stage as a crucial one in which responsibility aspects should be integrated and considered. Furthermore, from an outcomes perspective, innovation and R&D intensive firms, especially those whose customers are intermediaries and not the ones who will use the product or service, have not often considered how their innovations affect society. They may have looked at their sourcing and manufacturing process but paid less attention to what effects may be caused for people or the environment once their products or services are utilised. With the RI concept, these considerations of the effects on society and how to best serve to society enter the R&D functions in companies. The concept is also promising in addressing the core business responsibility for companies in sectors that are innovation-intensive.

While innovation itself has no normative orientation, some companies have made it part of their strategies or business models to innovate in order to reduce the negative impacts of their products or services on people and/or the environment. As an innovation management strategy, responsible innovation can be understood as a measure to reduce the risks of innovation failing to meet consumer wants, missing out on potential markets, or costly adaptations or roll-back at late points in the innovation process, while simultaneously increasing public credibility, legitimacy and trust of the company and its innovative products or services. All of which correspond to making innovative businesses more competitive.

The examples of AppNps and Yoti show that businesses can develop their own approaches and work out RI practices that suit them in their particular contexts.

---

[10] https://www.yoti.com/about/council/

Even though the two companies are vastly different, and the key issues of responsibility in their innovation processes are also different, both AppNps and Yoti have, (1) understood what RI in their specific context is and what are the most critical elements to address; (2) reflected on the potential benefits of RI for their specific case; (3) established management and employee commitment to RI; (4) established an action plan and roadmap to RI. The companies have in common a basic interest in creating positive societal impact. They both want to go beyond what is legally required of them in terms of safety, security or ethical issues. They want to play an active role in shaping the regulatory environment for their current and future business undertakings. Even though, like any other small enterprise, personnel time is among their scarcest resources, they invest time and effort into deliberation within and beyond company boundaries and into putting new practices and routines into place. Even though there are no numbers (yet) to predict a monetary return on these investments, both companies are committed to traveling this path and reaping economic profits based on the principles of responsible innovation.

Rather than implementing a top-down initiated policy concept, RI in companies concerns company values, innovation practices and interaction with consumers and other external stakeholders. If we accept that the RI concept is currently not tailored towards businesses, but that businesses are willing to implement elements of RI once they understand the benefits for their own business strategy as well as towards society, then a promising manner to approach implementation would be to build on what businesses are already doing – either individually or in sectoral initiatives – and provide them with information as well as with tools to explore other aspects of RI. Implementation strategies are highly likely to vary between sectors, application areas or even between individual businesses. Learning from "responsible innovation pioneers" among peers can constitute an important first step towards understanding how RI can be made operational in a business context.

References

Adams, R., Bessant, J., & Phelps, R. (2006). Innovation management measurement. A review. *International Journal of Management Reviews, 8*(1), 21–47.

Blok, V., & Lemmens, P. (2015). The emerging concept of responsible innovation. Three reasons why it is questionable and calls for a radical transformation of the concept of innovation. In B.-J. Koops, I. Oosterlaken, H. Romijn, T. Swierstra, & J. van den Hoven (Eds.), *Responsible innovation 2: Concepts, approaches, and applications* (pp. 19–35). Cham: Springer.

Brockhoff, K. (1992). R&D cooperation between firms—A perceived transaction cost perspective. *Management Science, 38*(4), 514–524.

Carroll, A. B., & Shabana, K. M. (2010). The business case for corporate social responsibility: A review of concepts, research and practice. *International Journal of Management Reviews, 12*(1), 85–105.

Chesbrough, H., Vanhaverbeke, W., & West, J. (Eds.). (2006). *Open innovation: Researching a new paradigm*. Oxford: Oxford University Press.

Crane, A., Palazzo, G., Spence, L. J., & Matten, D. (2014). Contesting the value of "creating shared value". *California Management Review, 56*(2), 130–153.

European Commission. (2011). Communication from the commission to the European parliament, the council, the European Economic and Social Committee of the Regions: A renewed EU strategy 2011–2014 for corporate social responsibility COM(2011) 681 final. Brussels.

Fichter, K. (2009). Innovation communities: The role of networks of promotors in open innovation. *R&D Management, 39*(4), 357–371.

Garcia, R., & Calantone, R. (2002). A critical look at technological innovation typology and innovativeness terminology: A literature review. *Journal of Product Innovation Management, 19*(2), 110–132.

Gassmann, O., & Enkel, E. (2004). *Towards a theory of open innovation: Three core process archetypes. Archetypes.* Conference paper. R&D management conference (RADMA), Lisbon.

Kieser, A. (1969). Innovationen. In E. Grochla (Ed.), *Handwörterbuch der Organisation* (pp. 741–750). Stuttgart: Poeschel.

Ma, M., & Morris, L. (2017). The agile innovation Sprint. *International Management Review, 13*(1), 92–98.

Moir, L. (2001). What do we mean by corporate social responsibility? *Corporate Governance, 1*(2), 16–22.

Moldaschl, M. (2010). Why innovation theories make no sense, Lehrstuhlpapiere // Professur für Innovationsforschung und Nachhaltiges Ressourcenmanagement, No. 9/2010, Professur für Innovationsforschung und Nachhaltiges Ressourcenmanagement. Chemnitz, Technische Universität Chemnitz.

Montiel, I. (2008). Corporate social responsibility and corporate sustainability: Separate pasts, common futures. *Organization & Environment, 21*(3), 245–269.

Montiel, I., & Delgado-Ceballos, J. (2014). Defining and measuring corporate sustainability: Are we there yet? *Organization & Environment, 27*(2), 1–27.

Phillips, R., & Caldwell, C. B. (2005). Value chain responsibility: A farewell to arm's length. *Business & Society Review, 110*(4), 345–370.

Porter, M. E., & Kramer, M. (2011). The big idea: Creating shared value. *Harvard Business Review,* January–February. https://hbr.org/2011/01/the-big-idea-creating-shared-value. Accessed 11 Oct 2018.

Roberts, E. B. (1987). *Generating technological innovation.* Oxford: Oxford University Press.

Schumpeter, J. A. (1939). *Business cycles: A theoretical, historical, and statistical analysis of the capitalist process.* New York: McGraw-Hill.

Utterback, J. M. (1996). *Mastering the dynamics of innovation.* Harvard: Harvard Business School Press.

Vedin, B. A. (1980). *Large company organization and radical product innovation.* Lund: Bratt-Institut für Neues Lernen.

von Hippel, E. (1986). Lead users: A source of novel product concepts. *Management Science, 32*(7), 791–805.

Chapter 3
Learning from Sustainability-Oriented Innovation

Katharina Jarmai

Abstract This chapter argues that insights from the realm of sustainability-oriented innovation can provide useful answers to the question of why Small and Medium-Sized Enterprises (SMEs) would (or should) become interested in implementing responsible innovation practices. It is based on the assumption that "responsible innovation" and "sustainability-oriented innovation" are different approaches aimed at orienting innovation towards increased positive impacts on social and natural environments. Motivations and influences for pursuing sustainability-oriented innovation have been studied in the past, and can provide insights into reasons for pursuing the implementation of responsible innovation practices.

Keywords Responsible innovation · Sustainability-oriented innovation · Corporate responsibility · Sustainable development · Corporate impact · Societal challenges

3.1 Introduction

Most research on responsible (research and) innovation has so far been conducted from a policy or socio-ethical perspective. In the early years of the debate, research on industry implementation was limited (Blok and Lemmens 2015; Blok et al. 2015; Lubberink et al. 2017). Recent years have, however, seen an increase in EU funding for analysing and supporting responsible innovation in industry. This has been accompanied by a growing number of peer reviewed papers investigating different aspects of responsible innovation in companies and industry sectors. EU projects have contributed to the implementation of responsible innovation in different types of organizations by means of tool kits, methods, self-assessment/self-check tools,

K. Jarmai (✉)
Institute for Managing Sustainability, WU Vienna University of Economics and Business, Vienna, Austria
e-mail: katharina.jarmai@wu.ac.at

© The Author(s) 2020   19
K. Jarmai (ed.), *Responsible Innovation*, SpringerBriefs in Research and
Innovation Governance, https://doi.org/10.1007/978-94-024-1720-3_3

training materials, etc. (Nwafor et al. 2017). However, company-specific tools that meet the realities of businesses are still missing. Furthermore, the crucial question of why companies should invest time and money into understanding and implementing tools for responsible innovation in the first place still remains to be resolved.

Innovation with the aim to decrease negative impact on the social and natural environment has been discussed for a couple of decades under the terms "eco-innovation", "environmental innovation" and "green innovation" (Schiederig et al. 2012; Díaz-García et al. 2015); and has been summed up under the term "sustainability-oriented innovation" (SOI) (Klewitz and Hansen 2014). These discourses have identified and classified a range of drivers and barriers for impact-oriented innovation. Applying the definition of sustainability-oriented innovation as "deliberate management of economic, social and ecological aspects" (Klewitz and Hansen 2014: 57) in innovation, we understand sustainability-oriented innovation and responsible innovation as two approaches aimed at orienting innovation towards increased positive impacts on social and natural environments.

The aim of this chapter is to explore sustainability-oriented innovation to identify similarities and differences between this kind of innovation and responsible innovation; and to learn from what has been found about what drives or impedes innovation that aims to increase positive impact on its social and/or natural environment.

3.2 Sustainability-Oriented Innovation

The concept of sustainability-oriented innovation has its roots in the notion of eco-innovation and the debate that followed publication of the *Brundtland Report* in 1987 (Klewitz and Hansen 2014). The *Brundtland Report* stated that "(…) the orientation of technology development must be changed to pay greater attention to environmental factors." (WCED 1987: para 65). It further pointed out that "Technologies are needed that produce 'social goods', such as improved air quality or increased product life, or that resolve problems normally outside the cost calculus of individual enterprises, such as the external costs of pollution or waste disposal." (WCED 1987: para 67). Since the 1990s, innovation with the aspiration to create positive environmental impacts has been studied under the terms "eco-innovation", "environmental innovation" and "green innovation" (Schiederig et al. 2012; Díaz-García et al. 2015). The debate has developed to include social criteria in addition to environmental ones; and has been carried forward under the terms "sustainable innovation", "sustainability-related innovation" and "sustainability-driven innovation" (Klewitz and Hansen 2014). The notion of sustainability-oriented innovation (SOI) subsumes these concepts to describe the integration of "deliberate management of economic, social and ecological aspects" (Klewitz and Hansen 2014: 57) in innovation.

Sustainability-oriented innovation can be defined as the commercial introduction of a new or improved product, service or system that leads to "environmental and (or) social benefits over the prior version's physical life-cycle" (Hansen and Grosse-Dunker 2013: 2407). In other words, sustainability-oriented innovations can be understood as innovations that replace less sustainable solutions on the market. Whether a new solution counts as a sustainability-oriented innovation thus depends on alternative options on the market. This relational character of sustainability-oriented innovation (Schaltegger et al. 2012, 2016) is also reflected in the understanding of sustainable entrepreneurship as transformative process (Adams et al. 2016).

Sustainability-oriented innovations can be differentiated into the categories used to describe different types of "regular" innovation with no normative requirements (see Chap. 2); i.e. product innovation (goods and services), process innovation (production or delivery method), marketing innovation (design, packaging, placement, promotion, pricing), and organisational innovation (business practices, workplace organization, external relations). In addition, increasing the service content of products can be considered another type of sustainability-oriented innovation. By increasing the service content of an innovation, its value for the consumer is decoupled from the amount of physical resources needed to produce it. Hansen and Grosse-Dunker (2013) describe three such product-service combinations: Adding a service to an initial product (e.g. a take-back service), product rental or leasing instead of sale (e.g. a car share service), or selling a result instead of a product (e.g. laundered clothes instead of washing machines). In this way, sustainability-oriented innovation can fulfil the same function, or meet the same needs, as an option that is already on the market but with an alternative, more sustainable solution. Companies may also develop sustainability-oriented innovation in an effort to go beyond fulfilling existing consumer needs and come up with entirely different solutions that encourage a more sustainable lifestyle.

Hansen and Grosse-Dunker (2013) identify five phases in the life-cycle of a product (supply chain, production, packaging/distribution, use, and end-of-life phase) in which positive impact can be created through sustainability-oriented innovation, and they provide examples of positive impact on the economy, environment and society; such as e.g. increased customer satisfaction, energy-efficient production, and safe and fair labour conditions. They emphasize the fact that the valuation of impact may change over time, as has been the case, for example, in the assessment of bio-fuels: "Although a short hype around its potential to fuel cars with renewable resources emerged, the enthusiasm for bio fuels was rather short-lived as the necessary cultivation of oil-bearing trees also implicated a displacement of food crops and thus negative side effects to the local population (e.g., advances of food prices)." (Hansen and Grosse-Dunker 2013: 2409).

3.3   Sustainability Versus Innovation?

Before an innovation reaches the market a succession of decisions are taken. This includes decisions about following up on a particular idea, and including or dismissing specific features of the novel product or process as well as design and marketing aspects. All companies are forced to balance their expenses with their revenues and to plan their investments based on expected returns. Small or Medium-sized Enterprises (SMEs) are typically particularly dependent on external (market) developments and thus tend to operate at relatively high levels of uncertainty. They need good reasons to allocate resources (personnel as well as financial) to activities that do not immediately support their core business model. This is particularly true when the expected return on their investment remains elusive.

In the discussion about sustainability-oriented innovation, three broad positions can be distinguished on the relationship between sustainable development and innovation (Fichter et al. 2006):

1. Ecologic and ethical considerations can hinder innovation;
2. Deteriorating environmental quality increases the pressure to innovate;
3. The guiding principles of sustainable development generate ideas and are a source of competitive advantage.

The literature on sustainability-oriented innovation has identified a number of barriers that impede the integration of sustainability criteria in innovation practices and strategies. Madrid-Guijarro et al. (2009), for example, list aspects connected to costs, risks, resistance to change, and difficulties in attracting and retaining qualified personnel (internal barriers), and to lack of information about markets or technologies and the need for additional support from the government or external partners (external barriers). In addition, small companies tend to have specialized portfolios and little access to venture capital. In addition to limited personnel and financial resources, sustainability-oriented innovation is further hindered by limited knowledge of decision-makers about two crucial aspects: First, their options to increase environmental and societal benefits, and second, the medium- to long-term benefits they can expect from doing so (Walker et al. 2008). This problem is sometimes exacerbated by information about companies' options to increase environmental and societal benefits that is inadequate in a business context, uses language which is too technical or academic, or is simply difficult to access (cf. Walker et al. 2008).

In the spectrum between highly formalised structures and decision-making processes and ad hoc decision-making, smaller and younger companies tend to appear at the ad hoc end. A lack of management and organisational structures as well as little planning of innovation processes can hinder innovation in SMEs. In environments that are defined by high speed, creativity and enthusiasm, the requirements posed by responsible innovation might inspire a fear of being slowed down and confined by forms, checklists and other bureaucratic obstacles.

3.4   Drivers for Sustainability-Oriented Innovation Practices

The literature on sustainability-oriented innovations identifies several potential drivers for the integration of sustainability criteria in companies' innovation strategies and practices. These drivers are traditionally based in innovation theory and environmental policy, and can be classified into supply-side factors, demand-side factors and the regulatory framework. Supply-side factors include technological and managerial capabilities and tangible and intangible assets, as well as knowledge and skills that enable companies to develop sustainability-oriented innovations. Collaborations with research institutes, private or public agencies and universities are also acknowledged as important sources of external knowledge. Demand-side factors include market demand and the way the company is perceived by its main target groups of customers. The regulatory framework includes laws, regulations and standards, such as those developed by the Eco-Management and Audit Scheme (EMAS) or the International Organization for Standardization (ISO), and is considered an important driver for the implementation of sustainability-oriented innovation in businesses. In a similar manner, Kesidou and Demirel (2012) differentiate between demand-side factors, organisational capabilities and the regulatory framework.

The multi-impulse model developed by Fichter (2005) is one prominent explanatory approach of a company's internal and external factors pushing or pulling innovation towards sustainability. The model is based on Schumpeter's (1947) model of creative response and understands innovation as a result of the creative performance of actors under specific framework conditions in which a combination of factors exert influence on the innovation process. The multi-impulse model illustrates company-internal (company vision, key individuals) and company-external influences (technological developments, market demand, regulation, civil society) on an innovation process. Fichter (2005) describes a range of company-internal and company-external factors that influence decisions and interaction in a company's innovation process. These factors are not isolated but can reinforce one another to affect the general orientation of an innovation process as well as particular decisions that are taken within it. Company-external factors include impulses through radical technological innovation, market demands, regulation and support mechanisms, public opinion conveyed via the civil society or the media, and national or sectoral overarching goals. How these external impulses are processed within a particular organisation depends on internal structures, actor constellations both within the organisation and with external actors, influential individuals, and internalised strategic orientation as well as basic cultural and value-based settings. Different studies emphasise the relevance of different internal or external factors. While Kopfmüller et al. (2001), for example, emphasize the role of technological developments, market demand and regulation in the context of sustainability innovation, Fichter et al. (2006, 2007) stress the importance of company culture and the intrinsic motivation of key individuals as decisive influences.

A variety of business motivations for conducting sustainability-oriented innovation have been described in the literature. The spectrum ranges from moral and ethical obligations, which evolve around the morality of products and services, their effects on human beings and social issues within global value chains, to economic motivations. Fichter et al. (2006) provide the following possible reasons for including considerations about impact on society and the environment in the innovation process:

1. Sustainable development is perceived as relevant by the company;
2. Sustainability-oriented visions serve as drivers for innovation in the company's relevant environments;
3. Changing legal frameworks require adjustments or adaptation;
4. Public funding is increasingly oriented by criteria of sustainable development;
5. Sustainable development becomes an important criterion for the financial market;
6. Prevention of reputation and acceptance losses.

According to Gil et al. (2001), competitive motivations positively relate to the implementation of environmental practices in a company. Opportunities to improve productivity or to reduce costs are expected from changes of business processes and products. In the absence of strong external push or pull factors, internal factors become more relevant in the decision to orient innovation processes and output to sustainability or responsibility criteria.

Dijkema et al. (2006) describe engagement with sustainability as process in four phases: In the first phase, a company primarily reacts to external pressure, forcing it to decrease negative impacts on its environment. The second phase is characterized by internal engagement with the topic of sustainable development, including discussions and the development of action strategies; this phase includes the adaption of innovation strategies. In phase three, the company commits to a more long-term orientation along the lines of sustainable development and takes appropriate measures such as the institutionalisation of new processes. Companies that have entered the fourth phase have advanced to formulating their own sustainable development strategies through interaction with external actors; they understand sustainable development as a continuous process, where innovation activities are continuously re-defined and re-formulated.

Other authors have differentiated indifferent, defensive, offensive and innovative (Steger 1993), or reactive, anticipatory and innovation-based (Noci and Verganti 1999) company strategies with regard to sustainability. What all of these definitions have in common is the distinction between intrinsic company motivation to assume responsibility for their actions towards their – social and/or environmental – environment, and external push and pull factors that increase pressure on the company to comply with certain laws, regulations, needs or expectations.

In recent years, the debate about sustainability-oriented business strategies has moved towards defining a business case for sustainability (e.g. Schaltegger and

Wagner 2006), which can create added value for the company through a range of aspects such as risk reduction, cost efficiency, reputational effects, market differentiation or market development.

3.5 Similarities and Differences Between Sustainability-Oriented Innovation and Responsible Innovation Practices

By comparing the definition of sustainability-oriented innovation quoted above to the definition of responsible (research and) innovation by von Schomberg (2011) as "transparent, interactive process by which societal actors and innovators become mutually responsive to each other with a view to the (ethical) acceptability, sustainability and societal desirability of the innovation process and its marketable products (…)", one major similarity and one crucial difference present themselves. First, both concepts concern the potential for delivering societal and ecological benefits through innovation; in the case of sustainability-oriented innovation through replacing a less sustainable solution already on the market, and in the case of responsible innovation through responding to sustainability demands and societal needs through innovation. Second, sustainability-oriented innovation addresses the improved performance of the innovative product, service or process, while responsible innovation focuses primarily on the research and innovation process and only secondly addresses the outcomes of the innovation process. While sustainability-oriented innovation is thus defined by its effect on the environment and/or by the intention of the innovator (Díaz-García et al. 2015), responsible innovation is mostly defined by innovation process qualities and criteria.[1] These matters have been recognized in the current debate on responsible innovation, and some recent definitions such as Sutcliffe (2018: 1), put more weight on the output qualities of innovation: "The concept of Responsible Innovation aims to focus attention on ensuring innovation delivers benefits to society; negative impacts are better anticipated and managed in advance and the involvement of people is important in shaping innovation."

The implementation of both sustainability-oriented innovation practices and responsible innovation practices require willingness, capacities and the development of capabilities to deal with diverse knowledge about economic, social and

---

[1] The European Commission has been promoting responsible research and innovation by funding projects on the thematic elements of ethics, gender and diversity, public engagement, open access, and science education through the previous and current European Framework Programmes "FP7" and "Horizon 2020". In the academic debate, a common agreement about key aspects of RRI has developed in the form of four dimensions that would lead towards more responsible innovation processes, entailing a collective and continuous commitment to conduct research and innovation processes in an anticipatory, reflective, inclusive (deliberative), and responsive way (Owen et al. 2013).

ecological contexts and phenomena. In addition, the decision to orient company strategy to sustainability or responsibility criteria may require a fundamental shift in mind set, from simply adhering to laws and regulations to actively creating a positive impact on the society and/or the environment. Engaging with sustainability or responsibility issues makes it necessary to gather and process knowledge from external sources. In the case of responsible innovation these external sources explicitly need to include civil society organisations (CSOs) and groups which are potentially put at a disadvantage.

3.5.1 Company Benefits

Both concepts work at the interface of business and society relationships, and are therefore confronted with the question of whether – and if so, which – benefits can be expected for businesses, or for society. In the context of responsible innovation it is frequently asked why companies would engage with the concept and invest time and resources into implementing corresponding practices. It is also a question for those who see it as their task to promote responsible innovation practices in companies or other research and innovation actors. How to best communicate what responsible innovation is all about? What expectations should be raised about benefits and added value? Is it advisable to try and "sell" responsible innovation as a "door opener" to hitherto unrecognized market segments and business opportunities? Is it naïve to focus on the potential of innovation actors to increase their positive impact on society through implementing responsible innovation practices?

Two of the positions about the relationship between sustainable development and innovation introduced by Fichter et al. (2006); i.e. that ecologic and ethical considerations can hinder innovation or, on the contrary, generate ideas and be a source of competitive advantage; are applicable to the context of responsible innovation. The remaining position, that deteriorating environmental quality increases the pressure to innovate is more difficult to interpret in the context of responsible innovation; and this highlights a critical difference between sustainability-innovation and responsible innovation. Sustainable development is driven by an acute need for change – even though the global dimension of this need might be difficult to grasp, it is still not being felt by large shares of the world's population and remains disputed by many. On a smaller scale, however, ecological deterioration can be felt in terms of reductions in air quality, extreme droughts or floods, disappearance of essential food components or similar local phenomena that impact particular communities. It seems fairly logical for companies to come up with better technological or other forms of solutions that will, for example, aid improving air quality in their customers' or employees' living environments. In comparison to these kinds of impacts, the added value of making sure that your innovation is "responsible" is much less obvious. There are a few reasons for this, which can be usefully discussed by focusing on the different constituting elements of responsible innovation:

- Anticipation – Similar to the more complex issues in sustainable development, the added value of taking responsibility for potential future applications of your product or the reciprocal effects of your process component is difficult to integrate into company strategies;
- Inclusion – While it has been said above that adding an inclusive element into company research and innovation can open the door to new market segments, action and change is necessary to profit from this opportunity. While the example about improving air quality should show that improving living conditions for a company's target groups is fairly obvious, investing extra effort into figuring out other groups potentially affected by a company's actions and taking measures to improve their living conditions is not;
- Ethics – Being more sustainable can often be in line with being more cost effective; e.g. by reducing the amount of material included in a product or the amount of energy needed in a production process. In contrast, being more responsible tends to be more cost intensive rather than less; at least on short-term time scale. Agreeing upon standards of ethical and responsible conduct in research and innovation, consulting with external ethics advisors or staying up-to-date on the latest data security regulations requires commitment, skills and time.

3.5.2 Assessment of Added Value

Another distinctive difference between what is defined as sustainability-oriented innovation and what has become known as responsible innovation lies in the possibility to assess the qualities of the final output and compare them to their less sustainable/less responsible alternatives on the market. Hansen and Grosse-Dunker (2013: 2407) define sustainability-oriented innovation as "the commercial introduction of a new (or improved) product (service), product-service system, or pure service which – based on a traceable (qualitative or quantitative) comparative analysis – leads to environmental and (or) social benefits over the prior version's physical life-cycle ('from cradle to grave')". With this definition, they recognize the possibility of applying methods such as lifecycle analysis or material flow analysis to compare two innovative products on the market that satisfy the same customer need. More than just recognizing this, they make this feature a constituting factor of what defines sustainability-oriented innovation. In the realm of responsible innovation, no comparable methods exist that would allow us to measure and compare the "responsibility" of two solutions on the market. There are at least two reasons for this: First, the level of "responsibility" might only become apparent at an unknown point of time in the future, whereas the amount of material or energy needed to produce both products can be compared even before the product goes on the market. Second, assessment criteria for responsibility – as defined in the responsible innovation concept – are difficult to capture in quantitative measures, whereas it is comparatively easy to calculate the amount of waste, for example, that is generated during the production of a product or by offering a particular service. This difference might be largely due to the fact that environmental criteria are central to sustainability-oriented innovation but not to responsible innovation. Proactive sustainability behavior can be measured through quantifiable values such as increased waste prevention measures or reduced material use (Klewitz and Hansen 2014).

In a literature review of 84 articles on sustainability-oriented innovation, Klewitz and Hansen (2014) identify innovation practices in SMEs and find that a total of 13 out of 20 identified practices can be allocated to the environmental dimension of sustainable development, while only seven concern non-environmental aspects such as stakeholder management, employee development and training or health and safety.

While sustainability-oriented innovation is defined in relational terms through increasing the sustainability of an innovative solution, responsible innovation has been discussed in terms of having been implemented or not (yet) in a particular organisation or network. Von Schomberg (2013) recounts an early distinction between responsible and irresponsible actors with regards to innovation and discusses examples of irresponsible innovation. Responsible innovation is still often treated dichotomously – it has either already been implemented in a particular organisation or sector, or it has not (yet) been implemented.

3.5.3 Individual Responsibility and Actor Networks

In the way that responsible (research and) innovation has been encouraged by the European Commission in its seventh and eighth Framework Programmes for Research and Innovation, until recently the addressees have been individual organizations. In the seventh Framework Programme (2007–2013), the focus was set on universities and research organizations. In the eighth Framework Programme (2014–2020), small, medium and large companies were included in the group of actors that were supposed to implement and promote responsible innovation practices. Many of these European projects have created tools to support their target groups in the implementation of responsible innovation. Such tools include management tools, a toolkit of activities and guidelines for engaging teenagers in STEM,[2] web 2.0 tools,[3] a toolkit for the design of public engagement activities,[4] tools for international cooperation, or the Gender-Diversity-Index (GDI).[5] All of these tools are targeted at individual organizations from academia and industry and provide them with customized, targeted support in the implementation of responsible innovation practices. Sustainability-oriented innovation however, goes beyond

---

[2] http://www.expecteverything.eu/hypatia/toolkit/

[3] http://nanopinion.archiv.zsi.at/en/about-nano/multimedia-repository.html

[4] https://toolkit.pe2020.eu/

[5] https://www.gedii.eu/wp-content/uploads/D3.1GenderDiversityIndex_final.pdf

the individual organisation by creating more sustainable production methods, market structures or consumption patterns (cf. Klewitz and Hansen 2014).

3.6 Complementing Responsible Innovation by Learning from Sustainability-Oriented Innovation

What can we learn from research on sustainability-oriented innovation to better understand company engagement with responsible innovation? The previous sections describe how the two concepts are sufficiently similar to assume similar categories of drivers and barriers in companies; even though characteristics and weightings will vary between industries and regions.

3.6.1 Intrinsic Motivation of Key Actors

Based on the first studies of the implementation of responsible innovation in companies, the intrinsic motivation of key company personnel and the strategic orientation of leading actors in the company's environment can be expected to be of similarly high relevance in the realm of responsible innovation. A recent analysis of expert interviews with Chief Executive Officers (CEOs) of SMEs in the Austrian medical device sector, for example, suggests that the moral standards of high-level decision-makers in companies are relevant when it comes to developing an overall innovation strategy (Auer and Jarmai 2018). In this sector, reasons to engage in innovation activities generally include profit-oriented elements, but also refer to the generation of positive impacts on customers, society, or the environment. Similar reasons have been documented for companies that develop eco-innovations or sustainability innovations and are often referred to as moral or intrinsic motivations (e.g. Clark and Charter 2007, Fichter et al. 2007). This suggests that while an expected increase in profits would likely be a good reason for companies to start looking into ways to implement responsible innovation, it does not have to be the only starting point. Moral motivations could open a second door to the implementation of responsible innovation in companies. Potential drivers and barriers to the implementation of responsible innovation are easily integrated into categories developed in the literature on sustainability-oriented innovation. As in the realm of sustainability-oriented innovation, all factors have the potential to act as either drivers or barriers, depending on other situational and contextual factors. Overall, the implementation of responsible innovation practices is more likely to be considered a benefit for a company if it is aligned with existing company practices and structures. Similar conclusions have been drawn from studies set in other sectors, such as ICT for ageing people, or the food industry (Chatfield et al. 2017; Blok et al. 2015).

3.6.2   Legal Frameworks and Public Funding

Adjustments or adaptation to changing legal frameworks and the orientation of public funding to responsibility criteria would fall under what Fichter (2005) describes as regulatory push and pull factors, which are probably the two most straightforward reasons for companies to engage with any kind of practice: Either because it is the law, or because they will only receive (public) funding if they comply with certain requirements. With regard to the different elements that make up responsible innovation, the ethical conduct of research and handling of sensitive information are probably the most highly regulated; at least in areas such as healthcare, through certifications and ethical compliance checks (Chatterji 2009). To the best of the author's knowledge, none of the other EU responsible research and innovation elements (European Commission 2012) or the process criteria described by Stilgoe et al. (2013) are currently required by law in any industry.

The European Commission has been promoting responsible (research and) innovation as a cross-cutting priority in the current Framework Programme for Research and Innovation "Horizon 2020"; this means that responsible innovation is not only the focus of particular research projects in the "Science with and for Society Programme", but that responsible innovation elements are also included as requirements for projects throughout the whole work programme. Within the Science with and for Society Programme, the European Commission has recently granted funding to a project with the objective to further integrate responsible research and innovation into research and innovation practice and funding at European, national and local levels.[6] The RRI-PRACTICE project[7] has previously conducted stakeholder workshops in 12 countries worldwide to assess the understanding of responsible innovation in national science, technology and innovation debates. Across these 12 workshops, "Awareness of the term RRI varied considerably across stakeholders, many having no prior knowledge of the term." (Owen et al. 2017: 1). At the same time, "Most institutions could readily identify national debates and ongoing activities related to responsible innovation framed as ethics, gender equality, public engagement and open access." (Owen et al. 2017: 2). At national level, the Netherlands Organisation for Scientific Research has had a Responsible Innovation programme since 2013. The UK Engineering and Physical Sciences Research Council (EPSRC) commits to ensuring that responsible innovation is "prominent in our strategic thinking and funding plans, including proposal assessment". The Research Council of Norway has implemented a 10 year programme dedicated to responsible innovation and CSR, with the primary objective to "address the grand global challenges through responsible technology development and socially responsible business organizations".[8]

---

[6] https://newhorrizon.eu/

[7] https://www.rri-practice.eu/

[8] https://www.forskningsradet.no/en/Funding/SAMANSVAR/1254004068509

3.6.3   Investors

Another potential push factor concerns the integration of selection criteria in the risk assessment and decision-making procedures of financial institutions. This also includes seed funding organisations and financial investors. Younger and smaller companies are particularly dependent on external funding sources to start a business in the first place or to cover costs of pursuing the development of a novel product or service. Similarly to sustainability criteria in financial investment, societal impact and responsibility are beginning to find their way into financial institutions. Black Rock, one of the largest global investment management corporations, state a commitment to "being a responsible corporate citizen and taking into account environmental, social and governance (ESG) issues". According to Black Rock, "… sustainable investing is becoming mainstream. Whether to mitigate risks, comply with regulation or target thematic impact, demand for these investment approaches has grown considerably."[9] First empirical findings suggest that financial investors who prioritise clients based on responsible innovation criteria would act as a driver for the integration of responsible innovation practices in SMEs (Auer and Jarmai 2018).

3.6.4   Company Reputation

A final potentially strong pull factor concerns the prevention of reputation and acceptance losses. This issue closely connects to the roots of the responsible research and innovation concept in the issue of public resistance to Genetically Modified Organisms (GMOs) in European society. In any innovation process, resources are invested without immediate return in the phase before the innovation goes to market, thus creating risk for the company. If money is invested in the development of an innovation and society rejects it immediately before or after it has entered the market, then the investment is lost. As Nathan (2015) argues, communication with societal actors before putting a final version of a product or service on the market opens up opportunities for adaption and re-consideration. Even though all of these options bring additional costs with them, any of them will be cheaper for a company than a complete roll-back after market entry. In addition, open communication of the objectives underlying a company's innovation can help to increase trust among target audiences; and it is considered that trust, often coupled with transparency, privacy or data security issues, will be among the most crucial business assets in the future (Leisinger 2017). SocietyInside has recently published a consultation document, "Principles for Responsible Innovation. For technologies

---

[9] https://www.blackrock.com/corporate/responsibility

society can trust",[10] which focuses the concept of responsible innovation around issues of trust and the trustworthiness of different research and innovation actors.

3.6.5 Combination of Supportive Policy Instruments

The literature on sustainability-oriented innovation describes different company-internal and company-external factors that influence the innovation decisions taken within an organisation. While some factors may be more inductive than others, decisions will generally be taken under the influence of a combination of company-external, company-internal, profit-oriented, mission-oriented and other factors. This means that no one adaptation in the company or its environment will single-handedly aid the implementation of responsible innovation, but that responsible innovation will likely be supported – like sustainability-oriented innovation – through a political pattern that combines different political instruments, creates economic incentives for the implementation of responsible innovation practices and provides orientation about funding requirements in the medium-term future (cf. Blazejczak et al. 1999).

3.7 Conclusions

This chapter describes the characteristics of sustainability-oriented innovation, relates them to responsible innovation and discusses potential reasons for pursuing the implementation of responsible innovation practices. While the European debate about responsible innovation originates in discourses on emerging technologies and research ethics and has been mainly driven by European research and innovation policy, the concept of sustainability-oriented innovation has its roots in the debate about technological progress for sustainable development and the production of social goods, and has been closely connected to corporate innovation management from the very early days.

Developing useful implementation options for a multifaceted, externally developed concept such as responsible innovation requires willingness, resources and structures to engage with the concept and integrate learning into company structures and practices. The requirements posed by responsible innovation might inspire a fear of being slowed down and confined by bureaucratic obstacles. This is particularly true when responsible innovation is communicated as policy regulation and in language that is too far removed from company realities for companies to easily grasp their substance. Legal frameworks, intrinsic motivation, easier access to financing and company reputation can counteract these challenges and function as drivers to the implementation of responsible innovation practices in companies.

---

[10] http://societyinside.com/our-principles-responsible-innovation

Legal requirements and conditions to receive funding are two of the most obvious reasons why companies will gear their innovation decisions towards a goal that does not immediately translate into economic benefit. If companies are forced to follow detailed ethical requirements in order to be able to bring a product to the market, for example the ISO 13485 certification concerning medical devices, they will make sure to complete the necessary procedures as quickly and thoroughly as possible.

Companies that consider the impact of their actions on the social and natural environment are often driven by strong intrinsic motivations. This suggests that while an expected increase in profits would likely be a good reason for companies to start looking into ways to implement responsible innovation, it does not have to be the only starting point. Moral motivations could open a second door to the implementation of responsible innovation in companies.

Overall, research to date on sustainability-oriented innovation suggests that innovation decisions in companies are influenced by a combination of company-external, company-internal, profit-oriented, mission-oriented and other factors. This means that a policy pattern aiming to support responsible innovation will need to combine different political instruments, create economic incentives for the implementation of responsible innovation practices and provide orientation about funding requirements in the medium-term future.

References

Adams, R., Jeanrenaud, S., Bessant, J., Denyer, D., & Overy, P. (2016). Sustainability-oriented innovation: A systematic review. *International Journal of Management Reviews, 18*(2), 180–205.

Auer, A., & Jarmai, K. (2018). Implementing responsible research and innovation practices in SMEs: Insights into drivers and barriers from the Austrian medical device sector. *Sustainability, 10*(1), 17.

Blazejczak, J., Edler, D., Hemmelskamp, J., & Jänicke, M. (1999). Umweltpolitik und Innovation. Politikmuster und Innovationswirkungen im internationalen Vergleich. In P. Klemmer (Ed.), *Innovation und Umwelt* (pp. 9–56). Berlin: Analytica.

Blok, V., & Lemmens, P. (2015). The emerging concept of responsible innovation. Three reasons why it is questionable and calls for a radical transformation of the concept of innovation. In B.-J. Koops, I. Oosterlaken, H. Romijn, T. Swierstra, & J. van den Hoven (Eds.), *Responsible innovation 2: Concepts, approaches, and applications* (pp. 19–35). Cham: Springer.

Blok, V., Hoffmans, L., & Wubben, E. F. M. (2015). Stakeholder engagement for responsible innovation in the private sector: Critical issues and management practices. *Journal on Chain and Network Science, 15*(2), 147–164.

Chatfield, K., Iatridis, K., Stahl, B. C., & Paspallis, N. (2017). Innovating responsibly in ICT for ageing: Drivers, obstacles and implementation. *Sustainability, 9*(6), 971.

Chatterji, A. K. (2009). Spawned with a silver spoon? Entrepreneurial performance and innovation in the medical device industry. *Strategic Management Journal, 30*(2), 185–206.

Clark, T., & Charter, M. (2007). *Sustainable innovation: Key Conclusions from Sustainable Innovation Conferences 2003–2006 organised by The Centre for Sustainable Design.* http://cfsd.org.uk/Sustainable%20Innovation/Sustainable_Innovation_report.pdf. Accessed 9 Oct 2018.

Díaz-García, C., González-Moreno, Á., & Sáez-Martínez, F. J. (2015). Eco-innovation: Insights from a literature review. *Innovations, 17*(1), 6–23.

Dijkema, G. P. J., Ferrão, P., Herder, P. M., & Heitor, M. (2006). Trends and opportunities framing innovation for sustainability in the learning society. *Technological Forecasting and Social Change, 73*(3), 215–227.

European Commission. (2012). *Responsible research and innovation—Europe's ability to respond to societal challenges.* Brussels, Belgium: European Commission Publications Office. https://ec.europa.eu/research/swafs/pdf/pub_rri/KI0214595ENC.pdf. Accessed 26 Sept 2018.

Fichter, K. (2005). Interpreneurship, Nachhaltigkeitsinnovationen in interaktiven Perspektiven unternehmerischen Handelns. Marburg, Metropolis.

Fichter, K., Noack, T., Beucker, S., Bierter, W., & Springer, S. (2006). *Nachhaltigkeitskonzepte für Innovationsprozesse.* Stuttgart: Fraunhofer-IRB-Verlag.

Fichter, K., Beucker, S., Noack, T., & Springer, S. (2007). *Entstehungspfade von Nachhaltigkeitsinnovationen. Fallstudien und Szenarien zu Einflussfaktoren, Schlüsselakteuren und Internetunterstützung.* Stuttgart: Fraunhofer IAO.

Gil, M.A., Jiménez, J.B., & Lorente, J.C. (2001). An analysis of environmental management, organizational context and performance of Spanish hotels. *Omega, 29*(6), 457–471.

Hansen, E. G., & Grosse-Dunker, F. (2013). Sustainability-oriented innovation. In *Encyclopedia of corporate social responsibility (2407–2417).* Berlin, Heidelberg: Springer.

Kesidou, E., & Demirel, P. (2012). On the drivers of eco-innovations: Empirical evidence from the UK. *Research Policy, 41*(5), 862–870.

Klewitz, J., & Hansen, E. G. (2014). Sustainability-oriented innovation of SMEs: A systematic review. *Journal of Cleaner Production, 65,* 57–75.

Kopfmüller, J., Brandl, V., Jörissen, J., Paetau, M., Banse, G., Coenen, R., & Grunwald, A. (2001). Nachhaltige Entwicklung integrativ betrachtet*: Konstitutive Elemente, Regeln, Indikatoren.* Berlin, edition sigma.

Leisinger, K. (2017). Responsible Research & Innovation: Science with and for society (with special consideration of the "leaving no one behind" aspect of the agenda 2030). Expert paper prepared for the H2020 project "COMPASS – Evidence and opportunities for responsible innovation in SMEs". https://innovation-compass.eu/wp-content/uploads/2018/01/Klaus-M.-Leisinger_RRI_Science-with-and-for-Society.pdf. Accessed 10 Aug 2018.

Lubberink, R., Blok, V., van Ophem, J., & Omta, O. (2017). Lessons for responsible innovation in the business context: A systematic literature review of responsible, social and sustainable innovation practices. *Sustainability, 9*(5), 721.

Madrid-Guijarro, A., Garcia, D., & Van Auken, H. (2009). Barriers to innovation among Spanish manufacturing SMEs. *Journal of Small Business Management, 47*(4), 465–488.

Nathan, G. (2015). Innovation process and ethics in technology: An approach to ethical (responsible) innovation governance. *Journal on Chain and Network Science, 15*(2), 119–134.

Noci, G., & Verganti, R. (1999). Managing 'green' product innovation in small firms. *R&D Management, 29*(1), 3–15.

Nwafor, C., Jarmai, K., Stacherl, B., & Montevecchi, F. (2017). *Integration of the RRI approach into collaborative R&D&I and SME participation in European funded collaborative research in healthcare, nanotechnology and ICT.* Benchmark Report and Policy Paper. Deliverable 1.4 of the Horizon 2020 project "COMPASS – Evidence and opportunities for responsible innovation in SMEs". https://innovation-compass.eu/wp-content/uploads/2017/09/D1.4-Benchmark-Report_Integration-of-the-RRI-approach-into-collaborative-Research-Development-Innovation-.pdf. Accessed 28 Oct 2017.

Owen, R., Stilgoe, J., Macnaghten, P., Gorman, M., Fisher, E., & Guston, D. (2013). A framework for responsible innovation. In R. Owen, J. Bessant, & M. Heintz (Eds.), *Responsible innovation, managing the responsible emergence of science and innovation in society* (pp. 27–50). Chichester: Wiley.

Owen, R., Ladikas, M., & Forsberg, E.M. (2017). Insights and reflections from National Responsible Research and Innovation Stakeholder Workshops. Report of the FP7 project "Responsible Research and Innovation in Practice" https://www.rri-practice.eu/wp-content/uploads/2017/09/Experiences-from-the-RRI-national-workshops-June-2017-final.pdf. Accessed 30 July 2018.

Schaltegger, S., & Wagner, M. (2006). Integrative management of sustainability performance, measurement and reporting. *International Journal of Accounting, Auditing and Performance Evaluation, 3*(1), 1–19.

Schaltegger, S., Lüdeke-Freund, F., & Hansen, E. G. (2012). Business cases for sustainability: The role of business model innovation for corporate sustainability. *International Journal Innovation and Sustainable Development, 6*(2), 95–119.

Schaltegger, S., Lüdeke-Freund, F., & Hansen, E. G. (2016). Business models for sustainability: A co-evolutionary analysis of sustainable entrepreneurship, innovation, and transformation. *Organization & Environment, 29*(3), 264–289.

Schiederig, T., Tietze, F., & Herstatt, C. (2012). Green innovation in technology and innovation management–an exploratory literature review. *R&D Management, 42*(2), 180–192.

Schumpeter, J. A. (1947). The creative response in economic history. *The Journal of Economic History, 7*(2), 149–159.

Steger, U. (1993). The greening of the board room: How German companies are dealing with environmental issues. In K. Fischer & J. Schot (Eds.), *Environmental strategies for industries* (pp. 147–166). Washington, DC: Island Press.

Stilgoe, J., Owen, R., & Macnaghten, P. (2013). Developing a framework for responsible innovation. *Research Policy, 42*(9), 1568–1580.

Sutcliffe, H. (2018). *Principles for Responsible Innovation. For technologies society can trust.* Consultation draft, Society Inside. http://societyinside.com/sites/default/files/Principles%20for%20Responsible%20Innovation%20Short%20February%202018_0.pdf. Accessed 10 Aug 2018.

von Schomberg, R. (2011). Towards Responsible Research and Innovation in the Information and Communication Technologies and Security Technologies Fields. SSRN Electronic Journal.

von Schomberg, R. (2013). A Vision of Responsible Research and Innovation. In R. Owen, J. Bessant, & M. Heintz (Eds.), *Responsible innovation, managing the responsible emergence of science and innovation in society* (pp. 51–74). Chichester: Wiley.

Walker, H., Sisto, L. D., & Mcbain, D. (2008). Drivers and barriers to environmental supply chain management: Lesson from the public and private sectors. *Journal of purchasing & supply management, 14*(1), 69–85.

WCED, S. W. S. (1987). World commission on environment and development. *Our common future.*

Chapter 4
What Responsible Businesses Can Learn from Social Innovation

Josephina Antoniou

Abstract This chapter presents initiatives and success stories from the realm of social innovation with the aim of identifying elements of Responsible Innovation (RI) and their significance. The motivation behind selecting social innovation to highlight the positive impact of RI practices is twofold. Focusing on social innovation provides, primarily, an opportunity to investigate the business perspective, by looking into cases where businesses have reconnected with the community through shifting their focus towards serving society, as a means to become more successful. Often this leads to immediate benefits for the business but also sets the framework for a long-term strategy that goes beyond well-known corporate social innovation activities, to encompass further activities that potentially initiate and support both social and environmental change. In addition, the focus on social innovation allows a better view of the community perspective, by considering the public as important business stakeholders, i.e. consumers and customers. As such, the public increasingly demands that business practices are handled in a more ethical way. As societies become more vulnerable due to economic instabilities, resource crises and political changes, the public demands adoption of new ways of thinking, and it is often implied that the road to a successful economic, and often cultural, transformation needs to go through social innovation. Undoubtedly, the goal of social innovation is to provide socially beneficial solutions that drive economic growth, but the task is not an easy one. Therefore, RI is essential for driving society forward, especially when it comes to the key aspects of employment, education and social inclusion.

Keywords Responsible innovation · Public good · Social innovation · Economic growth

J. Antoniou (✉)
University of Central Lancashire, School of Sciences, Pyla, Cyprus
e-mail: JAntoniou@uclan.ac.uk

© The Author(s) 2020
K. Jarmai (ed.), *Responsible Innovation*, SpringerBriefs in Research and Innovation Governance, https://doi.org/10.1007/978-94-024-1720-3_4

4.1   Introduction

This chapter aims to identify elements of Responsible Innovation (RI) in various initiatives and success stories from the realm of social innovation. The featured initiatives show that RI elements are significant in social innovation, especially in terms of public good.

The motivation behind selecting social innovation to highlight the positive impact of RI practices is twofold. Focusing on social innovation gives us an opportunity to investigate the business perspective, by looking into cases where businesses have reconnected with the community through shifting their focus towards serving society, as a means to become more prosperous and successful. Often this leads to immediate benefits for the business but also sets the framework for a long-term strategy that encompasses more than just the well-known corporate social innovation activities, but includes further activities that could potentially initiate and support both social and environmental change. We also focus on social innovation to open up the community perspective, by viewing the public as a group of important business stakeholders, i.e. consumers and customers. As such, the public increasingly demands that business practices are handled in a more ethical way. In fact, as societies are becoming more vulnerable due to economic instabilities, resources crises and political changes, the public demands adoption of new ways of thinking, and it is often implied that the road to a successful economic, and often cultural, transformation lies through social innovation.

The selected initiatives from business included in this chapter demonstrate how social innovation with RI elements have resulted in increased business productivity, lower business costs, and several other benefits. Section 4.2 presents the idea of innovation management through the development of new social incubators for socially innovative SMEs in Europe. Section 4.3 identifies RI elements in social innovation achieved through citizen-generated initiatives, while Sect. 4.4 follows social innovation and consequently RI aspects through addressing gender challenges. Section 4.5 offers the chapter conclusions.

4.1.1   Defining social innovation

Defining social innovation is challenging, especially within the scope of business growth. Looking at business growth through the lens of social innovation is often easier, especially in emerging markets "characterised by a rapidly expanding middle class and growing consumer spending" (Chakravorti and Siesfeld 2015). In such markets social innovation which stems from business includes activities that are community-engaging and inclusive, as well as activities that support business sustainability and further growth. In fact, social innovation aims at implementing effective solutions to important social, community and environmental issues, which are, more often than not, driven by local and global businesses in an effort to motivate societal growth and support business sustainability.

4.1.2 Social Innovation and Economic Growth

Undoubtedly, the goal of social innovation is to provide socially beneficial solutions that drive economic growth, but the task is not an easy one. RI as an outcome seeks to generate the right 'end points', which benefit people, planet, and profit (Sutcliffe 2011). Social innovation across Europe is such an 'end point', according to the Europe 2020 strategy, so it would seem necessary to identify the starting point in the values and rights of citizens of the European Society. To support these values and rights, the growth strategy for Europe 2020, articulates a vision for a smart, sustainable and inclusive economy, delivering high levels of employment, productivity and social cohesion (Sutcliffe 2011).

Innovation is essential for driving society forward, especially when it comes to the key aspects of employment, education and social inclusion. In order to move towards such benefits, it is important to involve the public and civil society stakeholders in the RI process. Thus, these stakeholders can become co-creators in the social innovation process. Given that RI aims to support a refinement of the roles and responsibilities of stakeholders with regards to innovation in business, specific elements of RI can strengthen social innovation.

In cases where the avenue for social innovation comes through business and, in particular, Small to Medium Enterprises (SMEs), then such innovation becomes more challenging. Overall, innovation is paramount to the survival and growth of any business (Mwangi and Namusonge 2014); however, SMEs are not always able to support RI fully. The incorporating of RI processes should be approached differently in SMEs to large organizations (which often support dedicated Corporate Social Responsibility (CSR) departments), because it may be the case that SMEs are unable or unwilling to consider the trade-off between immediate profit, and RI support mechanisms for non-immediate value in terms of profit.

RI elements in industry are usually aligned to a company's social innovation initiatives, even though RI discourse has "predominantly been designed to be applied to publicly funded research and innovation activities" (Soraker and Brey 2014). Even so, these activities are often linked with an intended social impact. To achieve the same social impact from privately funded initiatives, i.e. SMEs and larger organizations, this must be included in the planning stage for implementing social innovation initiatives, considering all stakeholders, in order to involve a range of expertise and perspectives (Stilgoe et al. 2013). Therefore, it becomes significant to focus on core aspects of RI and explore the applicability of these aspects in industry.

Consequently, this chapter identifies social innovation examples that demonstrate elements of RI and it highlights these elements, showing how they are aligned with industrial targets, and that RI implementation can often add benefits that can inform the business case. In the following sub-sections, a set of initiatives is presented and RI aspects are highlighted in corresponding social innovation success stories, to demonstrate the beneficial use of the particular RI aspects in industry, and to SMEs. The selected RI aspects include open innovation and open access and their contribution to business sustainability, environmental considerations, and ethical considerations, as well as policies, standards and codes of conduct, gender and diversity issues, and workplace equality.

4.2 Social Innovation Through Innovation Management and Incubation Initiatives

This section addresses social innovation through incubation, a popular mode of managing innovation and simultaneously promoting social impact. Innovation management is the management of such activities as idea generation, and the development of technologies, products or processes. Consequently, there is benefit in incorporating contributions from relevant experts, such as sociologists, psychologists, statisticians, engineers, etc. Innovation can be understood as novelty brought to (economic) use, so Research and Development (R&D) management can also be innovative. This has often been combined with CSR initiatives, and is now included through RI aspects in related industrial initiatives. Overall, it is important to understand that RI "is still very much a work in progress" (Soraker and Brey 2014). However, there is a need to move away from theoretical conceptualisation and consequent ambiguity, and offer more concrete translation of the RI concept into business practice (Owen et al. 2012).

RI can be a tool that provides a business with "guidance on how to move from an abstract concept to a more concrete approach" (Davalli 2017), so that entrepreneurs and business decision-makers can identify and take advantage of the new potential that RI can offer. This potential may vary between SMEs in different sectors, which can be explored through the exploration of specific challenging business sectors, e.g. where technology is key, such as biomedicine, nanotechnology, and cyber security. However, SMEs practicing in different sectors may also share common *horizontal* aspects of RI, which is particularly significant for SME innovation management.

For example, companies within the nanotechnology sector are expected to promote and support a high level of innovation that relates to society, as it is a sector "impacting modern social life and economies" (Galatsis et al. 2015). Recently the sector has been transformed by information technology. This transformation was caused by a rapidly growing technological sector, and has resulted in challenging social issues that need to be addressed, such as job losses, or gender equality and diversity in technical sectors. There is an imminent need for *responsible* practices to address these social issues, in addition to many others that fall under the umbrella of *responsibility*, such as environmental issues, or new policies, which are discussed in detail in subsequent sections.

Further examples of such aspects can be observed in the sectors of cyber security and biomedicine. Specifically, social innovation is addressed within the context of cyber security, where it is often related to the protection of social infrastructure from security threats. These new technological threats often relate to the increasingly popular *Internet of Things* concept, where devices and people are all connected, and are supported by the same networking infrastructure, with cyber security companies providing tools to protect this all-accessible infrastructure against *non-responsibly acting* malicious attackers (Miyao 2016). The cyber security effort needs to consider both the technological and organizational levels, and a wider implementation of RI can act as an ally in this effort. In addition, innovation and growth has been promoted within the area of biomedicine, which relates to "the changing relationship between the private and public sector in the use of human genomics and personal medical information" (Martin and Hollin 2014). The relationship is transforming into a collaborative one, offering a better foundation for *responsible* practices in the private sector that will be encouraged by the public sector. Martin and Hollin (2014) recognize that the sector is moving in this direction; "throughout the 2000s a series of UK and EU public policy initiatives were taken to promote innovation and growth of the [...] commercial development of biotechnology in particular".

According to Mwangi and Namusonge (2014), and based on Annual Reports from the United Nations Industrial Development Organization (UNIDO), SMEs constitute more than 90% of enterprises worldwide; therefore, it is significant for SMEs to have the opportunity to target cross-sectoral aspects and opportunities for added value. Overall, cross-sectoral aspects of innovation and responsibility have been targeted by CSR, a concept quite closely related to RI. However, the concept of RI is concerned with carrying out research and innovation responsibly, with consideration for the potential impacts for society; CSR is a more industry-driven concept, incorporating responsible strategies such as community philanthropies (Soraker et al. 2017), in order to strengthen businesses' profile, or corporations' roles in the market.

It is important to note that studies have shown CSR is associated mostly with large companies rather than SMEs. Larger companies are more concerned with their public profile because they attract more media attention and they are "particularly concerned to protect and enhance their reputation with the broader public as well as key stakeholders" (Smith 2013). Nevertheless, *responsibility* should concern companies of all sizes, especially when SMEs have obvious advantages, such as commitment by the management, personal relationships among employees, and less-obvious advantages such as comparatively fewer resources, all of which can help elevate business profitability potential by engaging with RI (Smith 2013).

The initiatives discussed in this section aim to demonstrate how RI aspects can be highlighted within social innovation case studies that have resulted in the following benefits: support structures for social innovation (e.g. incubators and additional job opportunities), increased business productivity and lower costs for the companies. In fact, incubation for social innovation startups can foster and support companies that are committed to innovation management structures, and consequently the adoption of several RI aspects.

The selected social innovation success stories begin with two European projects (BENISI and TRANSITION), which developed new social incubators for socially innovative SMEs. Both BENISI & TRANSITION are socially innovative projects launched by the European Commission that are mandated to help social entrepreneurship and innovation in Europe in local, regional and international contexts.

BENISI[1] is a project for building a European network of incubators for social innovation, where local social enterprises are supported in scaling up and growing beyond their locality. In addition to supporting the scaling up of social enterprises across Europe, TRANSITION[2] also provides learning output on which scaling methodologies are most effective in a given region. Statistics from the EBN Annual Observatory and Impact Hub Network on the success stories of BENISI and TRANSITION show the impact on social innovation across Europe. Specifically, incubated companies (i.e. socially innovative SMEs) show a high 90% survival rate after 3–5 years, post-incubation phase. For example in 2014, 3000 SMEs were developed in 150 incubators and innovation centres, creating 13,000 jobs with an average contribution by the SMEs of 8000 Euros per job (Davalli et al. 2016).

Another interesting initiative comes from Nigeria, although it was implemented in a very different setting from European SMEs. The Small and Medium scale industries Equity Investment Scheme (SMEIES) is an initiative of the bankers' committee in Nigeria. The commercial and merchant banks have agreed to set aside 10% annually of their profits before tax as their contribution to the development of SMEs, encouraging meaningful employment generation and the development of indigenous technology, (Mohammed and Abimiku 2015) to achieve significant social impact in terms of employment opportunities. These new financing approaches recognized the inherent weakness of SMEs in terms of resources, and the need to design finance schemes and products that are unique to them. To achieve the required employment generation through development of technology, innovation management, and more specifically responsible innovation, are inherently adopted by the specific companies.

Similar features and needs are observed to different degrees in SMEs globally. There are lessons to be extracted here, since one of the demotivating factors for SMEs when implementing RI is usually the lack of resources (Soraker and Brey, 2014). External support through incubation and further innovation management support could be sufficient motivation to engage SMEs with the RI implementation process, but external support need not always be financial; it can exist in the form of support for training (European Commission 2009), clustering (Organisation for Economic Co-operation and Development 2000), etc.

A third case study of social innovation through an incubation initiative is the case of Nesta in India; Nesta is an innovation foundation that looks for ways to bring

---

[1] www.benisi.eu

[2] www.transitionproject.eu

great ideas to life. To achieve this, Nesta invests in research and innovation, and manages this process by considering the difficulties of providing social impact in a challenging environment, since India is the home of a third of the world's poor (Gabriel et al. 2016). Nesta has succeeded in addressing this challenge by working with incubators to establish a more sophisticated set of impact and success metrics for incubation in low-income states in India. The success of Nesta comes from the fact that innovation management and specific aspects of RI (as per the previously presented definitions) are incorporated in the strategy to support business growth in the challenging environment, rather than primarily to achieve social impact, although it eventually does. According to Srikumar Misra, the Founder and CEO of Nesta, "We don't classify ourselves as a social enterprise… Our philosophy is about 'conscious capitalism'… Impact is built into the way we structured the business" (Gabriel et al. 2016).

Innovation management, and in particular corporate innovation management, as is the case when considering the implementation of RI aspects, has been addressed within many corporate strategies to provide a solution for the challenge of any corporate business; that the workforce is eventually trapped within a certain routine without ever thinking outside the box (Mitra 2016). Corporate innovation management aims to support ways of encouraging creativity and innovation. Sramana Mitra (2016), the founder of One Million by One Million, a virtual incubator helping one million entrepreneurs, globally, reach one million dollars, has repeatedly emphasized the importance of corporate innovation management and how the workforce should be trained to identify and validate innovation opportunities.

RI is closely linked to this concept, as it is a process, "by which societal actors and innovators become mutually responsive to each other with a view to the acceptability, sustainability and societal desirability of the innovation process" (Von Schomberg 2012). Implementing an RI framework within a company, whether this is a large corporation or an SME, ensures the desired corporate innovation management that will provide a platform for evolution and creativity through innovation; ensuring that it is explored *responsibly*, i.e. ethically and transparently in sustainable, highly reputable ways.

4.3   Social Innovation Through Citizen-Motivated Initiatives

Often, citizens are eager to participate in the social innovation process, as active stakeholders. Evolving communication technology is shifting the dynamics of the provider-customer relationship, offering the opportunity to customers, i.e. citizens themselves, to support (or not) a particular business and its practices. Companies can use open innovation tools to generate new ideas and invite the world to solve problems together, but also, consumers can act around an issue and pressure companies to change their behaviour.

Several examples of the use of social media in relation to responsible practice, which can strengthen or weaken a company's reputation, and consequently the potential for its future business sustainability, are presented in this chapter. One of them is the case of Crayola, which was featured in social media in 2013, as a company needing to enhance its recycling practices for marker pens. This campaign was started by an elementary school's class video, which got the attention of the company. Crayola responded by committing to helping reduce waste and pollution (Savedge 2013).

Such citizen-motivated initiatives have often developed from a local effort to a global campaign. Although there is a risk for companies when engaging in a dialogue with their customers, there are also a number of benefits; the main benefit being that by engaging with open communication tools, e.g. social media, an enhanced reputation of the business with its customer base may be achieved. For instance, according to *The Guardian* (2014), Unilever, which is a leader in CSR, has used an online open forum to hold a discussion for 2 years on sustainability issues, including the impact on the environment from the use of its products, and has posted a list of "challenges and wants" to request ideas for solving big issues, receiving more than one thousand ideas from the public in return.[3] Also, InnoCentive, EMC and EDF[4] ran its first *Eco-Challenge* activity in 2012 through similar platforms, seeking solutions for tracking shipments of used electronic components and subsystems to ensure that they are disposed of in a responsible manner. Another notable open innovation model included Heineken's $10,000 sustainable packaging contest[5] in 2012, which asked the public to propose ideas for sustainable beer packaging, and yielded numerous quality ideas from designers around the world based mainly on innovativeness and feasibility.

Consumers can act around a green issue and pressure companies to change their behaviour. Some notable campaigns have challenged specific companies, such as in the case of Crayola mentioned above. This provides an opportunity for companies to take action in order to rectify the relationship with their customers as well as to reinforce a *responsible* reputation once the social media movements break out. Other than Crayola, companies affected by such campaigns include Universal Pictures, and Dunkin' Donuts.

Universal Pictures was singled out by the public on social media for adding environmental education to *The Lorax movie, especially the movie website. The campaign was started by students and was completed successfully in 2012.* Dunkin' Donuts was asked by the public to replace its styrofoam cups, with a more environmentally-friendly alternative, by starting a petition that was eventually signed by more than three hundred thousand people, which gained recognition in national USA news. The company committed to the replace all styrofoam cups by 2020.

---

[3] https://www.greenbiz.com/blog/2012/06/07/how-unilever-crowdsourced-creativity-meet-sustainability-goals

[4] https://www.innocentive.com/emc-edf-and-innocentive-launch-new-eco-challenge-for-crowdsourced-solutions-to-key-e-waste-issue/

[5] http://www.sustainablebrands.com/news_and_views/articles/heineken-launches-open-innovation-challenge

4.4   Social Innovation Through Open Access

A concept related to open communication is open access. Open access is a research principle which promotes openness, transparency, and integrity. It aims to allow access to knowledge for all through open access to peer-reviewed literature, enabling the participation of society and improving research collaborations. This research and innovation principle appears at first glance to be contradictory to the private nature of enterprises participating in a competitive market, where it is not expected to share and collaborate. The goal is to find a common ground between the two worlds, in a way that will be beneficial for companies to adopt the principle of openness, without compromising their competitive edge in the market.

Where companies receive funding, e.g. through the European Commission's Horizon 2020 programme, then they must participate in open access publications. Moreover, initiatives such as Science 2.0 (trademark by ION Publications LLC), as well as concepts such as Open Science and Open Data, and Business Communication 2.0 that complement Open Access and motivate this paradigm further. In the following paragraphs, the aforementioned concepts will be further elaborated and contextualized.

The definition for the concept of Open Data is taken from the *Open Data Handbook* promoted by *Open Knowledge International,* a global non-profit organization focused on realizing open data's value to society. According to the *Open Data Handbook,* which is a set of guides, case studies and resources mainly for government and civil society on why and how to use open data, data must be both *technically open* and *legally open.* More importantly, the question of how Open Data is useful to SMEs is addressed in *Open Data: A twenty-first century asset for small and medium-sized enterprises* (Verhulst and Caplan 2015). The report presents 354 case studies of companies (SMEs and start-ups) that are using open data and how this can contribute to each company's economic growth. The report emphasizes that the open and shared data trend has "the power to fuel economic growth, job creation and new business opportunities".

The concept of Open Science promote scholarly sharing assisted by technology, especially new technologies like *Web 2.0.* The Science 2.0 initiative is based on the Open Science concept. Examples may include scientists using collaborative technology to share ideas, data or findings. Collaborative web technology provides several tools to achieve such collaborations, for instance, wikis, blogs and video journals (Waldrop 2008). Overall, open innovation achieved through Science 2.0 and Open Science concepts and tools, and further supported by Web 2.0 technologies, offer new opportunities for collaboration, research and education through harnessing collective intelligence (Tacke 2010). This can be a powerful asset for SMEs, which are usually under-resourced in terms of a wide variety of scientific expertise. This is especially applicable in highly technical sectors and corresponding SMEs.

The concept of Business Communication 2.0 addresses how the process of communication in business has been affected by, and needs to be further adjusted to certain technology advancements that are rapidly becoming the new communication norms. Social media and social networking technologies are examples of such new communication norms, where the underlying principles of the new communication pathways focus on the lack of geographical boundaries, and a sense of community, openness, and transparency. The use of new communication and networking paradigms has already begun to affect how business is done. Given the continuous evolution of virtualization, contextualization, and novel ways of data mining in the information world, it should be expected that this information-sharing trend will continue to penetrate the business communication world, through the use of more media reaching more stakeholders.

4.5 Social Innovation Through Addressing Gender Challenges

RI involves gender equality as one core societal aspect, which also intersects with other dimensions of RI (Lindberg and Schiffbänker 2013). Examples of social inclusion tackle aspects of gender and diversity, specifically aimed towards workplace equality. Gender issues and workplace equality are addressed in an attempt to drive teams and organisations towards their full potential, especially in sectors where the imbalance and lack of diversity are more broadly evident, such as technical sectors. Overall, there have been many initiatives to overcome this lack of diversity in research and technology, and the effort is ongoing with several European projects (FP7, H2020) currently addressing these issues (EGERA,[6] FESTA,[7] GARCIA,[8] GENERA[9]).

This section examines gender aspects in social innovation examples as these are related to RI. Specifically it examines the current under-representation of women in research and innovation, as well as the benefits of moving towards workplace equality and how this can be applicable for SMEs, as it is especially difficult to promote both CSR and RI initiatives in SMEs because of their size, compared to larger corporations. It is important to state here that for RI, "inclusion of equality issues should not be limited on addressing gender" (Soraker et al. 2017). In addition, other diversity categories such as age, ethnicity or disability may be equally important factors for SMEs for equality reasons, but also for the quality of their research and innovation. Therefore, although the Section mostly focuses on gender aspects, SMEs as well as larger corporations should also look to issues such as age(ing), migration, ethnicity, religion, disabilities, and sexual orientation, which are part of the European non-discrimination and equality mainstreaming policy (European Commission 2012).

Thus, gender is for example an issue in open innovation, addressing the question of *'Who is participating in research?'*; in ethics, addressing the question of *'How is*

---

[6] http://www.egera.eu/

[7] http://www.festa-europa.eu/

[8] http://garciaproject.eu/

[9] http://genera-project.com/

the risk of discrimination tackled?'; and in policy, addressing the question of *'Whether equality standards are implemented'*.

World Wide Worx (2014) has published articles stating that *"women are key to SME success"*. There has been an increase in female entrepreneurship since 2002 but especially since 2009 (VanderBrug 2013), but, having emphasized the importance of innovation for SMEs, there has been a lack of women in the research and innovation process (Busolt and Kugele 2009), with women forming only 28% of the world's researchers (UNESCO 2015). According to ITC News (2016), with statistics generated from approximately 20 countries around the world, close to 40% of all SMEs are owned by women. In addition, recent trends have shifted attention towards cultivating gender-friendly workplace cultures.

One of the most important initiatives addressing gender and equality issues in companies, and especially SMEs, is the *United Nations 2030 Development Agenda*, which has been adopted by 193 countries and focuses on sustainable development goals, especially on the *"transformation of discriminatory norms and gender stereotypes"*. Among the goals of the agenda to be achieved by 2030 is the following:

Ensure that all men and women, in particular the poor and the vulnerable, have the equal right to economic resources, as well as basic services, ownership and control over land and other forms of property. (United Nations 2015)

This includes policies aiming to encourage an increase of women in SME participation in international trade. The *UN 2030 Development Agenda* discusses the need for enhancing women's economic education and training to support their equal rights and responsibilities, giving special attention to promoting the economic role of women the economy in general, and particularly in SMEs.

A 2015 article by the International Labour Organization (ILO), highlights the contributions of the SCORE project in contributing to gender equality in SMEs in developing countries. The article emphasizes that the "ILO has long supported the fight for equality in the world of work, through the development and promotion of labour standards, gender focused campaigns – such as the Women at Work Century Initiative" and that it is important how the SCORE project raises *"awareness of gender issues among managers and employees"* through training and support visits to SMEs.

Another successful initiative comes from the European Small Business Portal, which features success stories from SMEs across Europe. The stories are featured on the European Commission website and include an initiative from a Slovakian SME. Regionfemme is run by entrepreneur Luica Haquel, which provides consulting and training for women to start up their own businesses. By 2014, only 5 years after funding was received by the EU to launch Regionfemme, and 2 years after the participants concluded their trainings, Regionfemme had already made a difference, with 56 out of 107 participants opening their own businesses, thus increasing women entrepreneurs in Europe.

Additional initiatives include: *UKRD Group, Gendered Innovations, Yellow Window*, and *GenPORT. UKRD Group* is a multimedia company in the UK; diversity management is central to the company's human resource strategic goals, which include the development of an inclusive and integrated workforce. *Gendered*

Innovations is an ongoing website project which collects and documents case studies of research and innovation which highlights the relevance of gender issues. *Yellow Window* provides a toolkit (process model and checklist) for considering gender during the whole research and innovation process. *GenPORT* represents a community of practice concerning gender issues in science, technology and innovation.

The benefits from such initiatives are twofold, i.e. both for the business and for society overall. Specifically, in the above case studies, it has been observed that the adoption of gender, and overall workforce equality practices, has resulted in increased business opportunities, enhanced workforce morale, more opportunities for training and for employment, and overall improvement in productivity.

4.6   Chapter Conclusions

Overall, this chapter identifies and discusses social innovation initiatives and success stories highlighting elements of RI implementation in companies across Europe, and especially SMEs. The chapter also addresses ways of creating a conducive environment for innovation, keeping in mind that social innovation is an emerging practice and discipline for NGOs (Bond 2016). Moreover, investigating the RI core aspects, it is evident that they cannot be useful in all possible corporate scenarios in the same way, especially where SMEs are concerned. According to the literature, incentives for an SME should directly reflect profit, or profit potential, and thus Soraker et al. (2017) emphasize the need to demonstrate that RI can result in:

> strengthening links with customers and end-users, enhancing the company reputation, decreasing business risks and unintended consequences, strengthening public trust in the safety of products, adopting an environmentally friendly profile.

The chapter shows that while profit-oriented incentives are certainly viable, a business's contribution to solving societal challenges is also an overarching benefit, and businesses have engaged in several social innovation initiatives, or have structured their operations so as to result in some kind of social impact. This has been achieved in several ways, including innovation management through incubation, open communication with citizens and customers, as well as attempts to support overall workplace equality practices.

Furthermore, the chapter discusses aspects of RI that are aligned with the above targeted social innovation case studies, where benefits have been evident and tackled by relevant research and discourse. Each section discusses the importance of the specific aspect and provides a list of initiatives and success stories to demonstrate the beneficial use of the particular aspect in industry, and more specifically SMEs, where the most challenges exist in implementing these aspects.

The selected cases included in this chapter demonstrate how social innovation with RI elements can result in increased business productivity, lower business costs, and several other benefits. Such cases include innovation management through the development of new social incubators for socially innovative SMEs in Europe, as well as social innovation achieved through citizen-generated initiatives. Finally, the chapter does not ignore the importance of addressing gender aspects in business through examples that demonstrate how the gender aspect can promote RI in business. It is significant to highlight through these cases that RI is essential for driving society forward, especially when it comes to the key aspects of employment, education and social inclusion.

References

Bond. (2016). *An introduction to Social Innovation for NGOs*, Licensed under a Creative Commons Attribution-Non Commercial 4.0 International License, London, UK.

Busolt, U., & Kugele, K. (2009). The gender innovation and research productivity gap in Europe. *International Journal of Innovation and Sustainable Development, 4*(2–3), 109–122.

Chakravorti, B., & Siesfeld, T. (2015). Business growth for good: Why context matters. *Stanford Social Innovation Review*, (Jul. 15, 2015).

Davalli, C. (2017). *All scale innovation, RRI tools blog*. Posted: January 2017. https://blog.rri-tools.eu/-/all-scale-innovation

Davalli, C., et. al., eds. (2016). *Scaling Social Innovation*. BENISI – TRANSITION: experiences and first success stories of the two European networks of incubators for social innovation. Joint Report of BENISI and TRANSITION projects, DG Research and Innovation, FP7-CDRP-2013-INCUBATORS.

European Commission. (2009). Guide for Training in SMEs, Directorate-General for Employment, Social Affairs and Equal Opportunities, Unit F.3.

European Commission. (2012). Responsible research and innovation: Europe's ability to respond to societal challenges. *Research and Innovation Policy*. https://doi.org/10.2777/11739.

Galatsis, et. al. (2015). Nanoelectronics research gaps and recommendations, *IEEE* Society and Technology Magazine, June 2015, pp. 21–30.

ITC News. (2016 October 16). How does gender affect the participation of SMEs in International Trade? *Speech by ITC Executive Director Arancha Gonzalez,* Canada.

Lindberg, M., & Schiffbänker, H. (2013). Entry on gender and innovation. In E. G. Carayannis (Ed.), *Encyclopedia of creativity, invention, innovation and entrepreneurship*. New York: Springer.

Martin, P., & Hollin, G. (2014). *A new model of innovation in biomedicine?* http://nuffieldbioethics.org/wp-content/uploads/A-New-Model-of-Innovation_web.pdf

Mitra, S. (2016). *Corporate innovation management: Lessons learned, one million by one million blog*. Posted: July 2016, Accessed: December 2016.

Miyao, T. (2016). Protecting social infrastructure from creeping threats – robust security against attack is not achieved in a day, *Social Innovation Hub,* Hitachi. http://social-innovation.hitachi/eu/topics/security/index.html

Mohammed, A., & Abimiku, J. (2015). Innovation among small and medium Enterprises in Nigeria. *Journal of Resourcefulness and Distinction, 11*, 1.

Mwangi, M. S., & Namusonge, M. J. (2014). Influence of innovation on small and medium Enterprise (SME) growth – A case of garment manufacturing Industries in Nakuru County. *International Journal of Innovation Education and Research, 2*(5), 31–41.

Organisation for Economic Co-operation and Development. (2000). *Small and Medium-sized Enterprises: Local Strength, Global Reach,* OECD Observer Policy Brief June 2000.

Owen, R., Macnaghten, P., & Stilgoe, J. (2012). Responsible research and innovation: From science in society to science for society, with society. *Science and Public Policy, 39*(6), 751–760.

Savedge, J. (2013). *Kids convince Crayola to recycle markers*, Mother Nature Network: Family, http://www.mnn.com/family/family-activities/blogs/kids-convince-crayola-to-recycle-markers

Smith, C. N. (2013). When it comes to CSR, size matters, *INSEAD Knowledge,* FORBES. http://www.forbes.com/sites/insead/2013/08/14/when-it-comes-to-csr-size-matters/#50c069811b6f

Soraker, J. H., & Brey, P. A. E. (2014). Current Discourse of RRI, health, demographic change and wellbeing, *Responsible Industry Deliverable D1.1, 7th Framework Programme.*

Soraker, J. H., et. al. (2017). *Models of RRI in Industry,* Responsible Industry Deliverable D3.3, 7th Framework Programme.

Stilgoe, J., Owen, R., & Macnaghten, P. (2013). Developing a framework for responsible innovation. *Research Policy, 42*(9), 1568–1580.

Sutcliffe H. (2011). *A report on responsible research & innovation, matter.* https://ec.europa.eu/research/science-society/document_library/pdf_06/rri-report-hilary-sutcliffe_en.pdf

Tacke, O. (2010). *Open Science 2.0: How Research and Education Can Benefit from Open Innovaton and Web 2.0;* Advances in Intelligent and Soft Computing book series: "On Collective Intelligence", pp. 37–48.

UNESCO. (2015). Women in science: UIS fact sheet, *UNESCO Institute for Statistics,* November 2015, No. 34.

United Nations. (2015). Transforming our world: The 2030 agenda for sustainable development, *united nations sustainable development knowledge platform,* A/RES/70/1.

VanderBrug, J. (2013). The global rise of female entrepreneurs. *Harvard Business Review,* September 4, 2013. https://hbr.org/2013/09/global-rise-of-female-entrepreneurs. Accessed 2 May 2017.

Verhulst, S., & Caplan, R. (2015). Open data: A twenty-first century asset for small and medium-sized enterprises, *governance lab 2014,* Creative Commons Attribution 4.0 International License.

Von Schomberg, R. (2012). Prospects for technology assessment in a framework of responsible research and innovation. In *Technikfolgen abschätzen lehren* (pp. 39–61). Springer.

Waldrop, M. (2008). Science 2.0. *Scientific American, 298*(5), 68–73.

World Wide Worx. (2014). *Women are key to SME success. http://www.worldwideworx.com/women/*

Chapter 5
RI – A Drain on Company Resources or a Competitive Advantage?

Doris Schroeder

Abstract Responsible innovation (RI) is an approach to business that can both incur and save costs. Some company leaders are concerned that it is yet another administrative and financial burden on their commercial operations. Others can see its financial advantages, e.g. avoiding the development of products the market will not accept, or reducing costs through sustainability measures. Building on the corporate responsibility and management advice literature, this chapter indicates a number of areas where RI can create a competitive advantage for SMEs. Real life case studies provide examples of reduced costs, reputational gains, employee retention, faster market entry, access to previously unavailable stakeholders, higher acceptability of end products, and higher innovation potential through diverse employees. Success cannot be guaranteed, but the willingness of an SME to innovate in areas that have positive societal impact in addition to profits can bring business benefits.

Keywords Responsible innovation · Profits · Competitive advantage · Corporate responsibility

5.1 Introduction

There are many reasons to start a business. The UK's 'No. 1 starting a business resource' lists 10. Reason 6 states that: "It can be very profitable" (Akselberg 2018). However, actual survival rates of small businesses are low. In the UK, 40% of small businesses do not survive the first 5 years (Lobel 2016).

One of the main causes of business failures is cash flow problems (ibid). For new businesses, it is therefore essential to use funds wisely. Anything that looks like a cost without a benefit will be avoided, and for good reason; small businesses need

D. Schroeder (✉)
School of Law, University of Central Lancashire, Pyla, Cyprus
e-mail: DSchroeder@UCLAN.ac.uk

© The Author(s) 2020
K. Jarmai (ed.), *Responsible Innovation*, SpringerBriefs in Research and Innovation Governance, https://doi.org/10.1007/978-94-024-1720-3_5

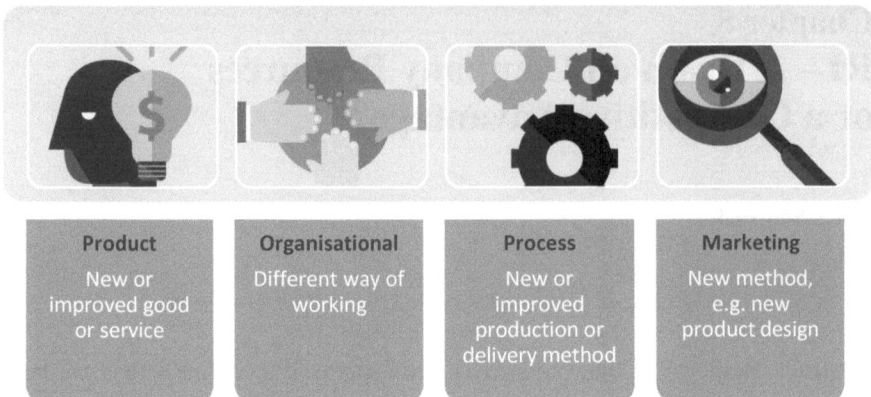

Fig. 5.1 Types of innovation

to concentrate even more than large corporations on spending money thriftily to avoid the cash flow trap.

Is the approach of Responsible Innovation (RI) a good business investment? As noted in previous chapters, RI invites researchers and innovators to engage with society to identify social and ethical impacts of the technologies they are developing, as well as to contribute innovative solutions to societal needs.

RI only applies to innovator companies. This means that RI will not be relevant for about half of European businesses. According to the latest Eurostat innovation statistics (Eurostat 2017), 49.1% of businesses reported innovation activity in the relevant period (2012–2014). The highest innovation levels were observed in Germany (67.0% of all enterprises), and the lowest in Romania (12.8%).

Four types of innovations are distinguished in these European statistics: product innovation, organisational innovation, process innovation and marketing innovation (see Fig. 5.1 and Sect. 2.2 in Chap. 2 of this book).

The following will introduce case study examples for each innovation type, which show how the application of RI principles can lead to a competitive advantage (either through savings or additional profits).

5.2 Product Innovation and RI

Innovative products is a term which can be used for both goods and services. Products have to be either entirely new to the market or a significant improvement on an earlier version. Typical examples of innovative goods are food, books, refrigerators, cars, computers, or fashion items. Typical examples of innovative services are flights, hotel nights, education, physiotherapy treatment or accountancy advice.

One can already see that innovation will be easier in some areas (e.g. cars) than in others (e.g. books). To provide focus for our question (is RI a good business investment?), it pays to ask about what happens in *ir*responsible innovation and the consequent impact on business.

Irresponsible innovation produces new products without due care for detrimental consequences:

> Our history is littered with the unintended consequences of innovations from destruction of stratospheric ozone by chlorofluorocarbons, to birth defects associated with thalidomide and mesothelioma associated with asbestos inhalation, to the near collapse of the global financial system in 2008, in which the innovation of complex financial products, such as the 'toxic' collateralized debt obligations… played no small part (Owen et al. 2013a).

As can be seen from this quote, innovation with highly detrimental consequences can occur both in goods (e.g. pharmaceutical products) *and* in services (e.g. finance). For example, as a result of the tragic birth defects and unnecessary deaths caused by the thalidomide disaster in many countries in the 1950s and 1960s, new rules for pharmaceutical testing and registration were issued, i.e. new responsibilities were added to existing procedures. This did not avoid the compensation payments required from the relevant companies, nor the massive reputational losses worldwide.

The near collapse of the global financial system in 2008 also had major detrimental effects on the industry worldwide. "Consumer trust in the sector plunged … [and] extensive regulation designed to clamp down on opaque investment banking practices, [hit] profits and [caused] a number of lenders to retreat from the sector"(Dunkley 2015).

Companies will want to avoid loss of life or well-being caused by product innovations for ethical reasons, but also for reasons of profitability. A promising angle for presenting the competitive advantages of applying RI is therefore the context of *risky* new goods or services. Risky can mean that disasters might occur, as in the above infamous scenarios, or it can mean that the product or service will not gain the trust of consumers and will therefore not be in demand, or may not receive official approval. An often cited example for the latter is the Dutch government's effort to move all patients' health records to an electronic system, an effort which was abandoned in 2011 due to major privacy concerns among citizens. At this point, 300 million Euros had already been invested (von Schomberg 2013). Whilst this failed investment will have been carried by the tax payer, a bad investment in an SME can reduce its survival chances significantly, given that one of the main causes of business failures is cash flow problems, as noted earlier. At the same time, Forbes count *lack* of investment as one of five reasons for businesses losing money (Kappel 2017). Hence, a balance between the two has to be found.

In this section the example of a technology perceived as risky is nanotechnology. The short case study will high-light a company, which invests in research and innovation whilst trying to reduce the chances that the market will reject the investment.

5.2.1 Product Innovation and RI in Nanotechnology

Nanotechnology is an enabling technology that has high rates of disapproval and distrust amongst the general public. Studies have found that the potential dangers of the technology are seen to override the potential benefits by many members of the public; in other words, according to a considerable section of the public, the technology has more disadvantages than advantages (Peter D. Hart Research Associates 2009). Using the food sector as an example, Vandermoere et al. (2011) write:

> In spite of great expectations about the potential of nanotechnology, this study shows that people are rather ambiguous and pessimistic about nanotechnology applications in the food domain.

As the European discussion around biotechnology has shown, high rates of pessimism cannot be aligned with commercially profitable product development using new technologies. For instance, the Court of Justice (2018) of the European Union ruled in July 2018 that organisms modified through new gene-editing tools are considered to be genetically modified and therefore fall under the 2001 GMO Directive). As a result, investment in gene editing and its product will be constrained in Europe.

Researchers and innovators have repeatedly expressed a concern that Europe will lag behind the United States and China in biotechnology developments, most recently after the ruling of the EU Court of Justice (Perets 2018).

The following section describes the case of a science-based spin off company, and its attempts to gain the public's trust for a nanotechnology product.

5.2.2 Nanotechnology Company applying RI

Responsible innovation in a business context (cross-ref to Chap. 2) introduced the company Applied Nanoparticles SL (AppNps) (Busquets-Fité et al. 2017) to readers of this book. AppNps main business is the commercial exploitation of a patent named BioGAS+. BioGAS+ uses iron nanoparticles to optimize anaerobic digestion processes. When added to organic waste it can increase the production of biogas. Renewable biogas can be used as a replacement for non-renewable natural gas and thereby contribute to sustainable energy use.

In contrast with genetically modified organisms, there are no specific regulations for nanotechnologies or nanomaterials at the European level. They fall under various other categories, for instance, Cosmetic Products, Novel Foods, or Medical Devices (EU Science Hub 2017). A *Code of Conduct for Responsible Nanosciences and Nanotechnologies Research* provides non-legally binding guidance (European Commission 2009).

AppNps decided to be pro-active about this legal uncertainty and focus on risk avoidance, which means employing safety by design approaches. According to several shareholders of AppNps:

It is well known that there are no specific regulations for nanotechnologies or nanomaterials at EU level. Instead, the manufacture, use and disposal of nanomaterials are covered, at least in principle, by a complex set of existing regulatory regimes… The consequence of this … is legal uncertainty. In the current legal framework and social context, companies need to develop safe and sustainable nanomaterials, and applying RI principles is the best way we found to achieve it (Busquets-Fité et al. 2017).

5.2.2.1 What Does That Mean in Practice?

AppNps has developed a vision which is supported by a tailor-made Code of Conduct. The Code includes articles about worker health and safety, as well as innovative articles, for instance about the relationship with suppliers, customers and society. Through continued engagement with relevant stakeholders, facilitated by the regular, transparent disclosure of information, the company anticipates providing a product that meets society's needs whilst generating a profit.

One problem AppNps shareholders have noticed with the diffusion of RI is that:

[E]stablished professionals often ... think that they are already 'responsible', and look at this [RI] movement with sympathy and condescendence, while young nanotechnology scientists … are more eager to adopt a responsible approach and realize that technology is never value-neutral, but always value-laden. They accept their moral responsibility (to critically reflect on the wider socio-ethical context of their work), and are thus ready to understand RI as a political tool. They only need the proper innovation environment (ibid.).

The AppNps shareholders believe that education of young scientists is the key to bringing RI into companies. The company is involved in educational efforts through their participation in EU-funded projects and writing up its experiences as, for example, in the case study summarized here.

5.2.2.2 Are Benefits in Evidence?

The 13 shareholders of the company believe that their vision, which incorporates a commitment to RI, helps them retain talented employees who might otherwise be easily head-hunted. The vision of the company is to make use of the opportunities that nanotechnology presents to generate wealth, but to pay special attention to sustainability and the minimization of deleterious side effects at the same time.

Responsible conduct of business operations is a theme that management consultancies increasingly promote for talent recruitment and retention. Forbes calls corporate social responsibility an underutilized asset:

Beyond benefits, compensation and work-life balance, there's an underutilized asset called corporate social responsibility, or CSR, that can attract and keep employees engaged at your company (Hattar 2018).

Deloitte believes that positive corporate impact, i.e. making the world a better place, can help recruit talented employees:

> As employees increasingly look for meaning and social impact in their corporate jobs, companies are seeking—and finding—ways to link talent development and rewarding, purpose-driven work, for both employee engagement and competitive advantage (Eggers et al. 2015).

AppNps can already state with confidence that an RI-linked approach to business helps retain talented employees. The company also hopes to gain the public's trust for its nanotechnology product through transparent information channels and public engagement. To date, it has not experienced resistance to the nanotechnology product BioGAS+.

5.3   Organisational Innovation and RI

Organisational changes involve different ways of working, for instance with new groups. Two organisational innovations will be introduced here, one leading to a product which has high appeal to the consumer through a university endorsement, and one which allows SMEs to adhere better to government requirements for stakeholder inclusion.

5.3.1   Collaborative Design for Biomechanical Devices

Universities are locations where cutting-edge research is undertaken. From the first-carbon positive houses to world-leading security systems for airports (*The Telegraph* n.d.), the groundwork and occasionally the implementation of many innovations is carried out by universities. At the same time, universities often suffer from a lack of impact of their ground-breaking research. In medicine, this implementation gap is described as the complex road from bench to bedside (Goldblatt and Lee 2010).

The term 'translational research' was created mostly by research funders to emphasize that even the best research has no impact on society if the theoretical know-how cannot be *translated* into helpful products and services (Woolf 2008). Universities have implemented various mechanisms to support start-ups, both within the institution (Houser 2014) and in the surrounding communities. One example is given in Box 5.1.

Earlier RI analysis involving SMEs have emphasized the importance of inclusion and end-user involvement (Stahl et al. 2017). If prospective end users are involved in product development, products can reach the market earlier and with higher acceptance levels, as shown in the success story of the ambiact presented in Box 5.2.

Box 5.1: Universities supporting start-ups – Northern lights, now propeller (https://propellerhub.co.uk/)
Northern lights/propeller is an enterprise incubator at the University of Central Lancashire. It started its operations in 2006 and has gone from strength to strength.

In 2013, Prime Minister David Cameron visited to meet young entrepreneurs and take part in a question and answer session with business people. He also announced a financial boost to the Government's Start-Up Loans scheme, which benefited the funding available through Northern Lights (UCLan Cyprus 2013).

In 2015, the Northern Lights Business Incubation Unit was awarded the title of Best Business Enabler of the Year at this year's Lancashire Business Awards (UCLan 2015).

The main offers of northern lights, now propeller, for early-stage innovators are 24/7 office facilities, sector-specific mentoring programs, links to local and national businesses, partnering events, financial and legal advice and the creative innovation zone (https://www.uclanfcci.co.uk/creative-innovation-zone.html).

Box 5.2: An RI success story – The ambiact (Frenken et al, 2018)
The ambiact is a smart meter for social alarm systems. It is designed as a plug-adapter and can be placed between the power outlet and any commonly-used appliance, such as a television, radio, or kettle. If not used for an unusually long time for the individual at risk, as previously defined, an alarm is raised. The developers of the ambiact included end-users throughout the entire design process in a co-design approach. Once the product had entered the market, they concluded:

Early engagement of stakeholders saves on costs: The engagement of future customers into the development process, starting during the initial idea phase and ending with cooperative product design, saved costs and time. The prototype itself was developed by adhering to acceptability factors for the customers. The continuous interviews during the field trials helped to identify potential problems, including around visual acceptability/impact of the product. Overall, the ambiact was developed from an initial idea to the final product in only three years with the involvement of a "work force" of volunteer end-users.

Fig. 5.2 Rehab Angel

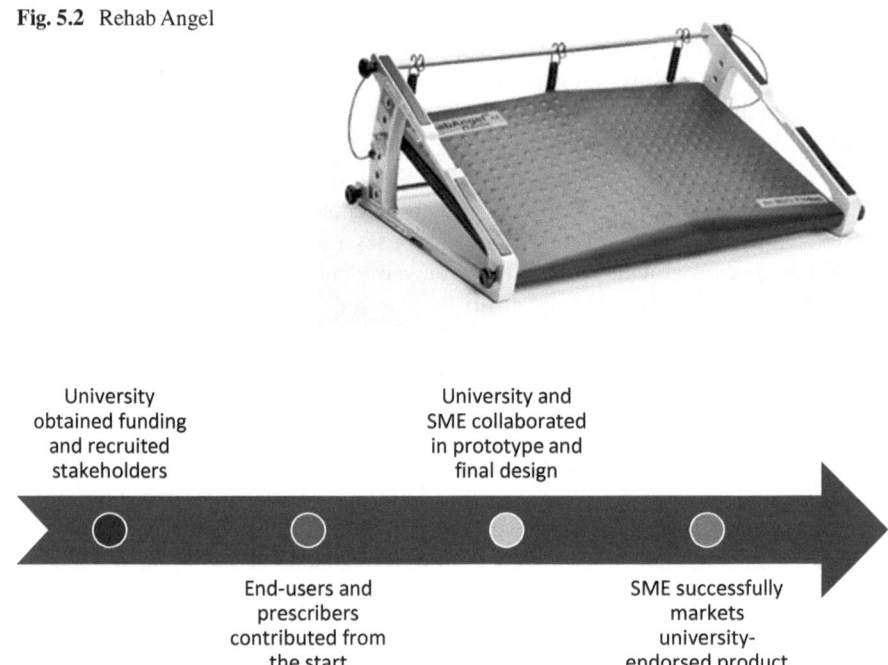

Fig. 5.3 Stakeholders involved in the Rehab Angel development

The case discussed here involves the co-operation of four stakeholder groups to create the Rehab Angel, an angled device used mostly for knee rehabilitation purposes (see Fig. 5.2). The inclusion of all stakeholders, including an SME, in product innovation led to benefits for all, which would have been much more difficult to achieve for each partner individually (Richards 2017).

The development of the Rehab Angel in the UK involved university researchers, an SME, end-users with knee problems as well as prescribers (e.g. physiotherapists). In combination these four groups achieved a result which was university-endorsed, yet business-marketed.

In an initial study, the university researchers identified a lack of evidence of the exact nature and 'dosage angle' of interventions used by clinicians. They then explored the use of squatting using decline boards and aimed to determine the optimum angle and the most effective regimen (Richards et al. 2008). As a result of initial publications, they were able to obtain prototype funding and recruit stakeholders, in particular end-users and prescribers, to the project. The SME joined the development early on and benefitted from the exposure to academic research.

This collaboration removed many of the latent issues around the innovation pathway, since key knowledge holders and product production systems were brought into the delivery of the project from the outset, see Fig. 5.3. At the end of the development circle, the SME was able to market the product successfully.

The benefits, including the commercial benefits, of this new collaborative way of working (through organisational innovation) are as follows.

The SME gained:

- access to cutting-edge scientific knowledge in their area of operations
- access to end-users who trusted the university's procedures to ensure safe studies with users
- university reports and independent peer reviewed papers to allow evidence-based marketing of the product

The university gained:

- societal impact, given that the SME successfully marketed the researched product, a new requirement for researchers at UK universities (Hefce 2014)
- the satisfaction that their ground-breaking research work will benefit patients
- new sources of co-funding through SME involvement in university activities

This shows the wider and longer-term benefits of broadening participation in a collaborative RI process between universities and businesses.

5.3.2   Widening the Work Force

Research and management literature from around the world shows that the involvement of women in the work force unlocks potential and improves performance (Joshi n.d.), (Devillard et al. 2016) A Gallup publication summarizes the reason for this effect very simply: "Men and women have different viewpoints, ideas, and market insights, which enables better problem solving."(Badal 2014). It is estimated that achieving 30% of women in leadership positions creates this effect (Heskett 2015). Studies have also shown that the increasing inclusion of staff from ethnic minorities in the workforce has positive impacts on profitability (Cox 2018).

As these are relatively uncontentious claims with significant existing media coverage and policy goals, the following focuses on more ambitious inclusion goals, namely the inclusion of disabled people in the work force of SMEs.

The "on my own" app (Vulterini 2018) was funded through an EU grant and developed in collaboration between socially responsible hotels and a range of partners. Its aim is to train people 'on the job' in the hospitality sector so that those with Down's Syndrome could work in the industry.

People with learning disabilities are harder to train on a job than those without, and they need more assistance and support to do a job well. To help trainers undertake this task, the app focuses on time management, work tasks and work tools, customised for each individual user. Pictures, videos and voice messages are used. The training profiles available are for: assistant cook, chambermaid, café waiter, restaurant waiter, breakfast waiter, receptionist, beach attendant and spa receptionist.

People with Down's Syndrome are, of course, theoretically able to undertake jobs in the hospitality industry, but previously this was not always practically feasible. For example people with Down's Syndrome often could not work autonomously, and required reminding of task content and timing on a regular basis. The "on my own" app is set up in such a way that a person trained on the job can – with the help of the app – undertake it with considerably less supervision and more autonomy.

Responsible innovation is about promoting diversity in the work force (European Commission n.d.-a), which this example clearly adheres to. However, the central question of this chapter is whether RI can create a competitive advantage for SMEs.

Hotels are almost always SMEs (with >10 and < 250 employees (European Commission n.d.-b) and in most European countries quota systems exist for the involvement of disabled employees.

> Quota systems for private and/or public enterprises or institutions can be found in the majority of EU countries (the exceptions are DK, EE, FI, LV, NL, SE and UK). Their basic target is to stimulate labour demand by committing employers to employ a certain share of employees with disabilities. Typically, the stipulated share ranges between 2% (ES) and 7% (IT) of the workforce. (Fuchs 2014)

It is no surprise that the app was designed under Italian leadership, given that the quota is highest in Italy (7%). This shows a problem-solving spirit under conditions of 'external pressure' (quota). Collaborating with academics and NGOs, as in this case, can reduce the industry costs of fulfilling the external requirement. As noted earlier, the advantage of using the "on my own" app for staff with learning disabilities is the significant reduction of supervisor time. This makes the app an example of where RI aligns with creating advantages for SMEs under conditions of requirements from the government. Costs (training and supervision) could be saved while a quota target is achieved.

5.4   Process Innovation and RI

A process innovation usually involves "a new or significantly improved production or delivery method" (OECD Glossary n.d.) which saves costs or increases consumer appeal or demand.

Responsible innovation is inextricably linked to sustainable-oriented research and innovation. Both are approaches which aim to increase the positive impact of innovation on society whilst minimizing the negative impact on the environment. Several theorists of RI, in particular Owen et al. and von Schomberg, believe that sustainability considerations should always be part of RI considerations.

According to René von Schomberg (2013), RI consists of three elements: ethical acceptability, sustainability and societal desirability. The most widely cited academic work on RI points to the necessity of respect for future generations. Sustainability is key to showing such respect. As Owen et al. write:

Responsible innovation is a collective commitment of care for the future through responsive stewardship of science and innovation in the present (Owen et al. 2013a, b).

Process innovations, which are focused on reducing negative environmental impact, are often called "green innovations", and defined as "new or modified processes, techniques, systems, and products to avoid or reduce environmental harm" (Marchi 2012). A successful case is given below.

5.4.1 *Mission Zero*

The highly ambitious "Mission Zero" of carpet manufacturer Interface is a good example (2017), with its mission introduced as early as the 1990s, for completion in 2020. Although Interface has over 3000 employees and operates on a global scale, it "operate[s] much like a small to medium-sized enterprise (SME) from an organizational and structural perspective."(ibid). For instance, Interface use a bottom-up management style without hierarchically and rigidly defined roles, and without formalized role training and career structure, thereby gaining time efficiencies. Its approach to green innovation therefore mirrors efforts undertaken by SMEs.

Two elements of this case study are important. First, the goals, and second, the innovation approach. The goals of Mission Zero are summarized in the following table (Table 5.1) (ibid).

The most important element of the innovation approach to Mission Zero was the setting up of the co-innovation team. The team uses the time and energy gained through the non-formalization of roles to encourage employees to "undertake discretionary activities above and beyond typical working practice, such as coming up

Table 5.1 Mission zero at interface

Mission zero goals	Description of goals
1. Eliminate waste	Eliminating waste in all forms – Material waste, wasted time and wasted effort
2. Benign emissions	Eliminating waste streams that have negative or toxic effects on natural systems
3. Renewable energy	Reducing energy demand and substituting fossil fuels with renewable ones like solar, wind and biogas
4. Closing the loop	Redesigning processes and products so that all resources used can be recovered at end of life and reused, closing the technical or natural loop
5. Resource efficient transportation	Transporting people with minimal waste and emissions. This includes consideration of plant location, logistics and commuting
6. Sensitising stakeholders	Creating a community within and around Interface that understands the functioning of natural systems and our impact on them
7. Redesign commerce	Redesigning commerce to focus on the delivery of service and value instead of material. Encouraging external organizations to create policies and market incentives

with new ideas, identifying resource needs, or reviewing established processes and products."(ibid) The authors of the case study commented that "the co-innovation team's utilization of organizational slack for innovative activity demonstrated one of the benefits of a high-slack environment for innovation", namely the time and energy available.[1] The most significant progress was made towards waste elimination (goal 1), benign emissions (2), renewable energy (3) and resource efficient transportation (5).

It was mostly waste reduction and energy savings, which led to the 2013 Interface statement that 480 million US dollars were saved through Mission Zero since 1994 (ibid.)

A project launched as part of Mission Zero in 2013 (Net-works) by Interface led a member of the co-innovation team to comment that "this project has greatly exceeded expectations, and it has gained an unexpected global recognition for its sustainability aspects." (ibid) This global recognition includes, to date, six prizes and awards, including the European Business Award for the Environment. According to the project's website (http://net-works.com):

> Net-Works™ redesigns global supply chains to create sustainable and scalable solutions that reduce marine plastic, increase fish stocks and improve the lives of marginalised coastal communities living in biodiversity hotspots of developing countries. We connect these communities to global brands via a fair and inclusive business model that delivers 'less plastic, more fish'.

Most of the Net-works operations are in the Philippines and Cameroon, with an expansion to Indonesia planned. From the first step (collecting discarded plastic fishing nets to avoid major marine pollution), the project expanded into setting up community savings and credit associations, and supply chains for seaweed carrageenan. The latter reduces over-fishing by creating a second means of income for local communities previously dependent entirely on fishing. The former applies the principles of fair trade and inclusive business to create livelihoods in disadvantaged communities.

This short case study shows how cost savings *and* global recognition can be achieved through green innovation by linking the aims of RI into a company's innovation processes.

Credit for the achievements of Interface has to be given to its founder Ray Anderson (2009), whose Ted Talk on the Business Logic of Sustainability succinctly makes the main points on the beneficial linkages between business advantages and sustainability. In the 12 years the Ted Talk account covers, net greenhouse gas emissions of Interface were down by 82%, sales increased by two-thirds, profits doubled and water usage decreased by 75%. These successes gave the company a "marketplace differentiator" (ibid.), and, Anderson says:

---

[1] The authors also noted that the lack of clear development paths meant that some employees "felt disengaged and underappreciated in their roles", ibid.

We have found Mission Zero to be incredibly good for business. A better business model, a better way to bigger profits. Here is the business case for sustainability. From real life experience, costs are down, not up, reflecting some 400 million dollars of avoided costs in pursuit of zero waste … And this dispels a myth too, this false choice between the environment and the economy. Our products are the best they've ever been, inspired by design for sustainability, an unexpected wellspring of innovation. Our people are galvanized around this shared higher purpose. You cannot beat it for attracting the best people and bringing them together. And the goodwill of the marketplace is astonishing. No amount of advertising, no clever marketing campaign, at any price, could have produced or created this much goodwill. Costs, products, people, marketplaces – what else is there? It is a better business model.

5.5 Marketing Innovation and RI

Marketing innovations change the look and feel of a product to achieve higher consumer ratings. The following case example is about packaging and is again related to sustainability. Packaging can be understood as a process innovation (an innovative change of the product's delivery method) or as a marketing innovation, given the strong importance of packaging for marketing a product.

The ways in which packaging can contribute to marketing has been explored by academics and in management circles. For instance, Rundh (2005) has advocated the use of packaging innovations to create competitive advantages. Vernuccio et al. (2010) argues that packaging is a tool beneficial to marketing, logistics, and ethics. Meanwhile, a variety of industries use packaging as a major marketing tool (Drinkpreneur 2016), and promote innovations for a range of reasons, for instance consumer convenience, product safety, or waste reduction, including cost savings (Cuneo 2017).

It would be beyond the scope of this chapter to explain in depth the complexity of sustainability solutions for packaging. This has been beautifully done in a TED Talk by Leyla Acaroglu (2013). Here, one example will be used, namely yoghurt pots, which are one of the most environmentally unfriendly forms of packaging. Yoghurt pots are single-serving foods and the small size and mixed materials make the single serving package highly unattractive for recycling (Wu 2014). The small size problem cannot easily be tackled in a time when the numbers of single person households are increasing considerably. In 2015, for the first time there were significantly more single-households in the EU than any other household type (Koessl 2017). However, some innovators are tackling the diversity of material mix, although solutions are difficult to find. Regarding sustainable packaging, *The Guardian* newspaper notes that "we were closer to an answer 30 years ago: what on earth happened to milkmen and bottle deposits? Now we live in an absurd age where a packet of crisps can have seven layers of wrapping" (Hall 2017).

German company Desto offers a light-weight, white plastic container for yoghurts, stabilized through a paper banderol, ready for separate recycling by 'green' consumers (Optipack n.d.). Whilst this may look like a success story, a sustainability expert in Germany explains that consumers do not seem to be separating the three components properly for recycling (paper, plastic, aluminium top), leading to a worse waste result than the traditional yoghurt pot. At the same time, people are buying the innovative pot in an effort to contribute to sustainable consumption (Lemke 2018).

What can be learned from this example regarding RI?

1. Consumers are buying innovative packaging which is designed to be more sustainable; there is a market.
2. Waste can be reduced through innovative packaging, e.g. the plastic content of the Desto container is significantly lower than the standard pot. As a result, material costs can be saved.
3. As noted above, employee attraction and retention can be improved with commitments to social goals. Sustainability is one of the top goals. Research found that employee pride, the perception that sustainable operations care for their employees and the link to personal value systems are the main reasons for this phenomenon (Network for Business Sustainability 2013).

There is clearly room for movement in the packaging sector, which is relevant to almost all product businesses. Sustainable packaging could achieve a considerable advantage for innovators whilst simultaneously achieving RI goals.

5.6   Conclusion

This chapter has described a range of business advantages for SMEs who engage in research and innovation responsibly. They are summarized in the Fig. 5.4.

Innovative SMEs have to decide for themselves. Do they want to innovative responsibly through their operations or not? This decision cannot be based solely on whether this makes businesses more profitable. Successful examples have been shown, but no guarantee is possible. The closest one can argue is that SMEs, which subscribe to responsible innovation, are more likely to attract and retain talented staff. Job satisfaction is higher when employees can be proud of their company. The rest depends on the context and on the willingness of the SME to innovate in areas that have positive societal impact in addition to profits. The possibilities are endless, from more sustainable packaging to reaping the benefits of an increasingly diverse work force. It is the initiative that counts and this has to come from SMEs themselves.

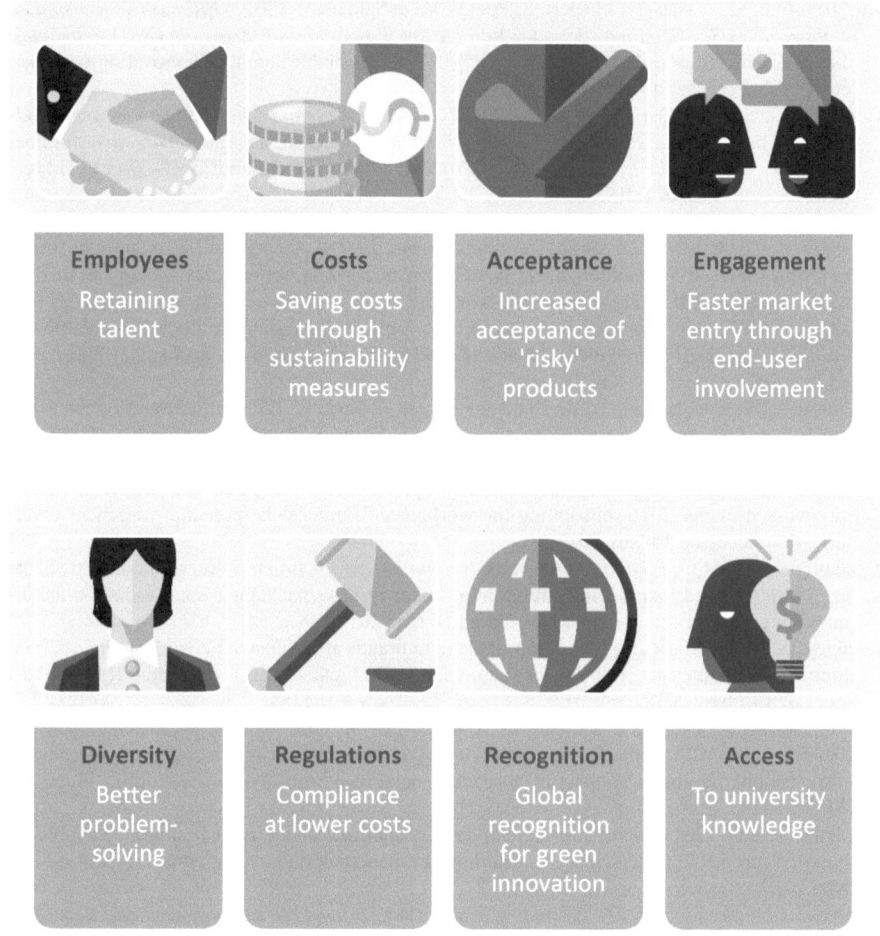

Fig. 5.4 Business advantages of implementing RI

References

Acaroglu, L. (2013). *Paper beats plastic? How to rethink environmental folklore,* [online] Ted Talks. Available at: http://www.ted.com/talks/leyla_acaroglu_paper_beats_plastic_how_to_rethink_environmental_folklore. Accessed 17 Nov 2018.

Akselberg, A. (2018). *10 reasons to start a business.* [online] Startups. Available at: http://startups.co.uk/10-reasons-to-start-a-business. Accessed 17 Nov. 2018.

Anderson, R. (2009). *The business logic of sustainability,* [online] Ted Talks. Available at: http://www.ted.com/talks/ray_anderson_on_the_business_logic_of_sustainability/transcript?language=en. Viewed 18 Nov 2018.

Badal, S. (2014). *The business benefits of gender diversity.* [online] Gallup Workplace. Available at.: http://www.gallup.com/workplace/236543/business-benefits-gender-diversity.aspx. Accessed 18 Nov 2018.

Busquets-Fité, M., Casals, E., Gispert, I., Puntes, V., & Saldaña, J. (2017) *RRI case study. Applied nanoparticles SL: Spinning off under Responsible Research and Innovation(RRI) Principles.* 1st ed. [pdf] Responsible Innovation COMPASS. Available at: http://innovation-compass. eu/1932-2. Accessed 17 Nov 2018.

Court of Justice of the European Union. (2018). *Organisms obtained by mutagenesis are GMOs and are, in principle, subject to the obligations laid down by the GMO Directive.* [online] Available at http://curia.europa.eu/jcms/upload/docs/application/pdf/2018-07/cp180111en. pdf. Accessed 17 Nov 2018.

Cox, J. (2018). *Businesses perform better when they have greater ethnic and gender diversity, study reveals. Independent* [online] Available at: http://www.independent.co.uk/news/business/ news/business-ethnic-gender-diversity-performance-levels-better-study-workplace-office-mckinsey-a8166601.html. Accessed 18 Nov 2018.

Cuneo, L. (2017). *Top innovations in packaging.* [online] Packaging Strategies. Available at.: http://www.packagingstrategies.com/articles/89355-top-innovations-in-packaging. Accessed 18 Nov 2018.

Devillard, S., Sancier-Sultan, S., de Zelicourt, A., & Kossoff, C. (2016). *Reinventing the workplace to unlock the potential of gender diversity.* Women Matter 2016. [online] McKinsey & Company. Available at.: http://www.mckinsey.com/~/media/mckinsey/featured%20insights/ women%20matter/reinventing%20the%20workplace%20for%20greater%20gender%20diversity/women-matter-2016-reinventing-the-workplace-to-unlock-the-potential-of-gender-diversity.ashx. Accessed 18 Nov 2018.

Drinkpreneur. (2016). *Packaging innovation is a major tool for brand owners.* [online] Available at: http://www.drinkpreneur.com/beverage-industry-news/packaging-innovation-is-a-major-marketing-tool-for-brand-owners. Accessed 18 Nov 2018.

Dunkley, E. (2015). Banks fight to repair damage to brands after financial crisis. *Financial Times* [online] Available at.: http://www.ft.com/content/bf47c4b4-ef22-11e4-87dc-00144feab7de. Accessed 17 Nov 2018.

Eggers, W. D., Wong, N., & Cooney, K. (2015). *The purpose- driven professional.* [online] Deloitte Insights. Available at.: http://www2.deloitte.com/insights/us/en/topics/corporate-responsibility/harnessing-impact-of-corporate-social-responsibility-on-talent.html. Accessed 14 Oct 2018.

EU Science Hub. (2017). *How are nanomaterials regulated in the EU?* [online]. Available at: http://ec.europa.eu/jrc/en/science-update/how-are-nanomaterials-regulated-eu. Accessed 17 Nov 2018.

European Commission. (2009). *A code of conduct for responsible nanosciences and nanotechnologies research.* [online] Available at.: http://ec.europa.eu/research/science-society/document_library/pdf_06/nanocode-apr09_en.pdf. Accessed 14 Oct 2018.

European Commission. (n.d.-a) *Horizon 2020, Science with and for Society* [online] Available at: http://ec.europa.eu/programmes/horizon2020/en/h2020-section/science-and-society. Accessed 18 Nov 2018.

European Commission. (n.d.-b) *What is an SME?* [online] Available at: http://ec.europa.eu/growth/ smes/business-friendly-environment/sme-definition_en Accessed 18 Nov 2018.

Eurostat. (2017). *Innovation statistics.* [online] Available at: http://ec.europa.eu/eurostat/statistics-explained/index.php/Innovation_statistics#Innovation_in_SME.E2.80.99s_and_in_large_ enterprises. Accessed 17 Nov 2018.

Frenken, T., Eckert, R., Jüptner, A., & Hein, A. (2018). AmbiAct. *ORBIT Journal*, [online] 1(3). Available at.: http://www.orbit-rri.org/ojs/index.php/orbit/article/view/68. Accessed 17 Nov 2018.

Fuchs, M. (2014). *Quota systems for disabled persons: Parameters, aspects, effectivity, European Centre for Social Welfare. Available at.* https://www.euro.centre.org/downloads/detail/1459. Accessed 17 Nov 2018.

Goldblatt, E.M., and Lee, W. (2010). From bench to bedside: the growing use of translational research in cancer medicine. *American Journal of Translational Research,* [online] 2(1), 1–18. Available at: http://www.ncbi.nlm.nih.gov/pmc/articles/PMC2826819. Accessed 17 Nov 2018.

Hall, D. (2017). Throwaway culture has spread packaging waste worldwide: here's what to do about it. *The Guardian* [online] Available at.: http://www.theguardian.com/environment/2017/mar/13/waste-plastic-food-packaging-recycling-throwaway-culture-dave-hall. Accessed 18 Nov 2018.

Hattar, M. (2018). *How to attract and retain top marketing talent through values. [online] Forbes. Available at.* http://www.forbes.com/sites/forbescommunicationscouncil/2018/07/20/how-to-attract-and-retain-top-marketing-talent-through-values/#45aaa8d9584b. Accessed 14 Oct 2018.

Hefce. (2014). *REF impact policy.* [online] Available at: http://webarchive.nationalarchives.gov.uk/20180319120009/http://www.hefce.ac.uk/rsrch/REFimpact. Accessed 18 Nov 2018.

Heskett, J. (2015). *Does gender diversity in management enhance performance? Why?* [online] Harvard Business School. Available at.: http://hbswk.hbs.edu/item/why-does-lack-of-gender-diversity-hurt-performance. Accessed 18 Nov. 2018].

Houser, C. (2014). *Why the university is the ideal startup platform. [online] Wired. Available at.* http://www.wired.com/insights/2014/02/university-ideal-startup-platform. Accessed 17 Nov 2018.

Joshi, R. (n.d.) Does gender diversity improve firm performance? Justjobsnetwork. Available at: https://ilo.org/wcmsp5/groups/public/%2D%2D-dgreports/%2D%2D-dcomm/documents/publication/wcms_616213.pdf. Accessed 17 Nov 2018.

Kappel, M. (2017). *5 causes for a small business losing money. [online] Forbes. Available at.* http://www.forbes.com/sites/mikekappel/2017/05/03/5-causes-for-a-small-business-losing-money/#5b956305c3df. Accessed 17 Nov 2018.

Koessl, G. (2017). *The rise of single households in the European Union and the impact on housing.* [online] Sociology Lens. Available at.: http://www.sociologylens.net/article-types/opinion/rise-single-households-european-union-impact-housing/18676. Accessed 18 Nov 2018.

Lemke, E. (2018). *The noble art of throwing away – how an organic yoghurt pot is becoming a problem[Blog] Eveline Lemke.* Available at.: http://www.eveline-lemke.com/2018/03/the-noble-art-of-throwing-away-how-an-organic-yoghurt-pot-is-becoming-a-problem. Accessed 18 Nov 2018.

Lobel, B. (2016). *Business failure: Four in ten small companies don't make it five years.* [online] Small Business. Available at.: http://smallbusiness.co.uk/business-failure-four-ten-small-companies-dont-make-five-years-2533988/. Accessed 17 Nov 2018.

Marchi, V. (2012). Environmental innovation and R&D cooperation: Empirical evidence from Spanish manufacturing firms. *Research Policy, 41,* 614–623.

Network for Business Sustainability. (2013). *Discover how good corporate citizenship and sustainable business practices attract and help retain top-notch talent for your firm.* [online] Available at: http://nbs.net/p/three-reasons-job-seekers-prefer-sustainable-companies-6f1780d2-0b1d-4778-aef9-20527ab78895. Accessed 18 Nov 2018.

OECD Glossary of Statistical Terms. (n.d.). *Process innovation.* Available at: http://stats.oecd.org/glossary/detail.asp?ID=6870. Accessed 18 Nov 2018.

Optipack. (n.d.). *Paper covering Desto.* [online] Available at: http://www.optipack.de/becher/desto_en.html. Accessed 18 Nov 2018.

Owen, R., Bessant, J., & Heintz, M. (2013a). *Responsible Innovation: Managing the responsible emergence of science and innovation in society* (1st ed., pp. xvii–xviii). London: Wiley.

Owen, R., Stilgoe, J., Macnaghten, P., Gorman, M., Fisher, E., & Guston, D. (2013b). A framework for responsible innovation. In R. Owen, J. Bessant, & M. Heintz (Eds.), *Responsible innovation: Managing the responsible emergence of science and innovation in society, 1st* (pp. 27–50). London: Wiley.

Perets, E. (2018). *Why does Europe lag behind the US and China in the gene editing race.*[online] Labiotech.eu. In *Available at.* http://labiotech.eu/features/gmo-gene-editing-europe. Accessed 17 Nov 2018.

Peter D Hart Research Associates, Inc. (2009). *Nanotechnology, synthetic biology, & public opinion.* 1st ed. [pdf] Washington, DC: Project on Emerging Nanotechnologies. Available at: http://www.nanotechproject.org/publications/archive/8286/. Accessed 17 Nov 2018.

Richards, J. (2017). *RRI case study the development of new rehabilitation devices for use in the community setting – The Rehab Angel.* Responsible Innovation COMPASS. Available at: https://innovation-compass.eu/wp-content/uploads/2017/11/Case-Study-3_Rehab-Angel.pdf. Accessed 18 Nov 2018.

Richards, J., Thewlis, D., Selfe, J., Cunningham, A., & Hayes, C. (2008). A biomechanical investigation of a single-limb squat: implications for lower extremity rehabilitation exercise. *Journal of athletic training*, [online] *43*(5), 477–82. Available at: http://www.ncbi.nlm.nih.gov/pmc/articles/PMC2547867. Accessed 17 Nov 2018.

Rundh, B. (2005). The multi-faceted dimension of packaging: Marketing logistic or marketing tool? *British Food Journal, 107*(9), 670–684.

Stahl, B.C., Obach, M., Yaghmaei, E., Ikonen, V., Chatfield, K., & Brem, A. (2017). The Responsible Research and Innovation (RRI) maturity model: Linking theory and practice. *Sustainability,* [online] *9*(6), 1036. Available at: http://www.mdpi.com/2071-1050/9/6/1036. Accessed 17 Nov 2018.

The Telegraph. (n.d.). *Ten Cutting Edge University Research Projects.* [online]. Available at: https://www.telegraph.co.uk/education/educationnews/12053483/Ten-cutting-edge-university-research-projects.html?frame=3529606. Accessed 17 Nov 2018.

UCLan. (2015). *UCLan business support service wins regional award.* [online] Available at: http://www.uclan.ac.uk/news/northern_lights_award_2015.php. Accessed 17 Nov 2018.

UCLan Cyprus. (2013). *Prime minister visits UCLan UK.* [online] Available at: http://www.uclancyprus.ac.cy/prime-minister-visits-uclan-uk. Accessed 17 Nov 2018.

Vandermoere, F., Blanchemanche, S., Bieberstein, A., Marette, S., & Roosen, J. (2011). The public understanding of nanotechnology in the food domain: the hidden role of views on science, technology, and nature. *Public Understanding of Science,* [online] *20*(2), 195–206. Available at: https://www.ncbi.nlm.nih.gov/pubmed/21657134. Accessed 17 Nov 2018.

Vernuccio, M., Cozzolino, A., & Michelini, L. (2010). An exploratory study of marketing, logistics, and ethics in packaging innovation. *European Journal of Innovation Management, 13*(3), 333–354.

von Schomberg, R. (2013). A Vision of RI. In R. Owen, J. Bessant, & M. Heintz (Eds.), *Responsible Innovation: Managing the Responsible Emergence of Science and Innovation in Society* (1st ed., pp. 51–74). London: Wiley.

Vulterini, P. (2018). *RRI case study: On my own…at work – a framework and an app.* 1st ed. [pdf] Vienna: Responsible Innovation Compass. Available at: http://innovation-compass.eu/wp-content/uploads/2018/01/Case-Study_4-On-my-own…at-work.pdf. Accessed 18 Nov 2018.

Woolf, S. H. (2008). The meaning of translational research and why it matters. *JAMA,* [online] *299*(2), 211–113. Available at http://jamanetwork.com/journals/jama/article-abstract/1149350. Accessed 17 Nov 2018.

Wu, A. (2014). Good product, bad package: top sustainable packaging mistakes. *The Guardian* [online] Available at: https://www.theguardian.com/sustainable-business/2014/jul/18/good-product-bad-package-plastic-recycle-mistakes. Accessed 18 Nov 2018.

Chapter 6
Engaging Small and Medium-Sized Enterprises in Responsible Innovation

Catherine Flick, Malcolm Fisk, and George Ogoh

Abstract A significant part of responsible innovation is engagement with diverse groups of stakeholders; this remains true for projects investigating responsible innovation practices. This chapter discusses strategies for engaging small and medium-sized enterprises (SMEs) in co-creating visions of and plans for implementing responsible innovation, drawing on the example of engagement with United Kingdom cyber security companies. The key aspect of the engagement was building trust between the responsible innovation researchers and the companies. Trust was built by a movement away from traditional recruitment procedures for research projects, towards proactive engagement with the culture and traditions of the sector – participating in company sponsored talks and conferences, finding ways to communicate effectively, and ensuring a tailored message that fit the expectations and requirements of the sector. This chapter reviews the context in which the recruitment took place, the assumptions made prior to recruitment, the approaches taken, the revisions made to these approaches, and ultimately offers some general recommendations for industry engagement in responsible innovation activities.

Keywords Cyber security · Responsible innovation · Engagement · Small-medium enterprises · Trust

6.1 Introduction

Some of the most significant challenges for responsible innovation in industry include raising awareness of the concept, showing businesses its value, and capturing businesses' interest in implementing responsible innovation in their own research and development practice. This chapter looks at the engagement of cyber security SMEs (small and medium-sized enterprises) in responsible innovation, by

C. Flick (✉) · M. Fisk · G. Ogoh
School of Computer Science and Informatics, De Montfort University, Leicester, UK
e-mail: cflick@dmu.ac.uk; malcolm.fisk@dmu.ac.uk; george.ogoh@dmu.ac.uk

© The Author(s) 2020
K. Jarmai (ed.), *Responsible Innovation*, SpringerBriefs in Research and Innovation Governance, https://doi.org/10.1007/978-94-024-1720-3_6

71

investigating the techniques that were used to recruit companies for a series of online and face-to-face peer co-creative workshops on implementing responsible innovation within the United Kingdom cyber security sector. The analysis of these engagement methods culminates in a set of general requirements and recommendations for engaging primarily with cyber security companies, but which also have general relevance to other industry sectors.

Responsible innovation, as has been seen in previous chapters, is a set of practices by which researchers and innovators engage with society to identify social and ethical impacts and issues of the technologies they are developing. Largely referred to in the academic world as the more cumbersome "responsible research and innovation" (RRI), definitions of responsible innovation are many and varied, but the general idea is that innovation should include society, deliberate on ethical and social issues, and align with societal needs (European Commission and Directorate-General for Research and Innovation 2013; Owen et al. 2013; Von Schomberg 2013). However, the concept of RRI does not have much penetration into industry (Stahl et al. 2017), and industry players are more likely to know and recognise terms such as corporate social responsibility (CSR) (European Commission 2011) or more simply, business ethics. To reflect this finding, and for reasons we discuss below, we henceforth refer to RRI as "responsible innovation" (RI).

In order to engage effectively with cyber security companies on topics surrounding RI and engage them in the planned workshops, a communications strategy needed to be devised. The approaches that were successful focused on the opportunities available to SMEs, were individually tailored to their spaces and requirements, and helped to ensured that the SMEs were comfortable in discussing confidential information. These experiences found that a desire for the development of trust with the general public, consumers of companies' products and services, and/or other businesses was a major driving factor in their engagement with RI.

This chapter reviews the context in which the recruitment of the companies took place, the assumptions made prior to recruitment, the recruitment approaches taken, the revisions made to these approaches, and offers some general recommendations for industry engagement. It argues that one of the most effective strategies for recruitment and engagement of SMEs is to become involved in the existing communication spaces of the sector, rather than expecting companies to respond to calls for interest.

6.2 Responsible Innovation for Cyber Security Companies

Previous chapters have explored the potential benefits of following a value-based approach to corporate innovation. However, the value propositions need to be well-defined and to generally align with existing goals within the company if they are to be considered useful. For example, for cyber security, trust seems to be a significant factor in interest in RI. The value that public and customer trust has for each cyber security company is significant, although this might not initially have been seen by cyber security SMEs in monetary terms and business sustainability. However, when RI activities were explained to cyber security SMEs in the context of ethics, responsibility, privacy, and trust, and with only a passing mention of 'responsible innovation' (instead of attempting to define RI explicitly), companies could see the alignment with their existing value statements, medium-long term goals, and discussions that had already taken place internally (especially regarding ethics). In fact, for cyber security companies, ethics and trust are regular topics of industry discussion, with philosophical differences arising between different camps on particular ethical dilemmas, such as disclosure of vulnerabilities (responsible disclosure vs. full disclosure), and bug bounties (rewards offered by companies for finding exploitable bugs in their software) (Hughes 2015; Lefkowitz 2017).

The emphasis on security is growing in a more uncertain and technologically-dependent world. Cyber security is therefore a natural growth area for industry, and a good example subsector of the more general IT industry, much of which grapples with uncertainty. It is a loosely defined sector encompassing many different types of security-related products and services. Much of the cyber security market is business-to-business, offering reputation protection, security of data, forensics and fraud detection, and server security. However, cyber security companies are also responsible for products and services that consumers use, such as security cameras, identity management apps, encryption of devices, and educational materials. The nature of cyber security's past can be suggestive of a somewhat 'cowboy' culture, with its frontiers of technological crime prevention often seen as a 'grey area' - including 'white-hat' (those who operate within legal and ethical norms), 'grey-hat' (those who operate mainly in a legal sphere, but occasionally exploit opportunities of policy vacuums, usually within ethical norms), and 'black-hat' hackers (those who break legal and ethical norms) operating on both sides of the law to meet their goals. Coupled with the complexity of the topic and issues, as well as poor representation of the field in movies and TV shows, there is a significant lack of understanding of what cyber security is, what its goals are, and how it works. This can translate into a lack of trust between end-users and security companies and their products, or to a view of cyber security products and services as 'grudge purchases' made by companies who view the sector much as they see insurance.

Thus, the value of RI to cyber security companies is in helping them to develop these trust relationships with their clients, whether they are individual end-users or companies. In this way, a company can show its trustworthiness to users who may not understand the technicalities, theoretical aspects, or even the user interfaces for cyber security. And in helping the cyber security sector to engage in openness, transparency, ethics and responsibility, along with other RI practices, clients who do not understand the inner workings of the technologies involved can develop a stronger trust relationship with the company.

We found in our work that companies are eager to engage with the concept of trust. The strategies detailed in the following section point to ways of harnessing companies' interest.

6.3   Recruitment Aims and Strategies

This section looks briefly at the aims of the RI research to provide context, then in more depth at the strategies which were chosen to approach the cyber security community to participate in the project, and describes the most effective engagement processes. It also discusses complications that arose after companies had made a commitment to the process. The resulting approach was effective in engaging companies for the RI workshops and helped build a significant rapport with the companies that, in turn, improved the outcomes of the workshops.

The aim of these interventions was to engage companies in a series of on- and off-line workshops. It was envisaged that the companies would work together to develop a 'responsible innovation roadmap' co-creatively with their peers, facilitated by the workshop leaders. Initially, there were to be three webinar-style online workshops, and two face-to-face workshops, where the companies would come together to co-create the shared roadmap using foresight and backcasting methodologies.[1] The cyber security companies were to be from the UK and considered as SMEs (up to 250 employees). SMEs were targeted as approximately 50% of SMEs in the UK are engaged in innovation activities (Department for Business, Energy, and Industrial Strategy 2017), but unlike large companies, they often do not have significant corporate social responsibility (or similar) arms.

Prior to the strategy being developed by which SMEs would be approached, however, a concern arose that cyber security companies might be more difficult to engage than other sectors due to their more secretive nature, particularly if this was to be in a peer-led co-creative exercise such as the planned workshops. This concern was based on discussions with cyber security experts within academia about company involvement with their research, but, as this article will show, the concerns were relatively unfounded, as the topics of ethics, trust, and other technical philosophical discussions were seen by the companies *individually* to be interesting and relevant. However, it took some time to realise this specific entry point for engaging with companies, as is explained below. The peer-led co-creative exercise however, was correctly identified to be a problematic approach for this sector, regardless of interest in the topics. The evolution of the planned co-creative exercises is also detailed below.[2]

Firstly, a generic, academic-style call for participation was developed. This was sent to a number of contacts identified by members of the research project. Some effort was made to circulate this call through established cyber security fora, for example, the UK Cyber Security Forum, as well as more personal networks, such as university cyber security partners. This was based on the assumption that companies would be most likely to respond to personal contacts and through advertisements on an industry website.

---

[1] More information on the workshop methodology and approaches can be found in D2.5 at https://innovation-compass.eu/deliverables-2/

[2] Examples of the drafts discussed below are available from the authors by request; due to space limitations we have included only the final, successful, recruitment letter in Appendix.

After poor engagement with this method (i.e. none), discussions with several cyber security experts were undertaken (university researchers with industry contacts; cyber security experts from other countries). Advice was taken on the nature of the 'sales pitch' (i.e. the description of the activities and the benefits to the companies in taking part) to make it more focused on the benefits that companies might gain from participating, as well as to avoid the implication that this would be largely an academic activity that might berate companies for unethical behaviour. The 'RRI' terminology was removed at this stage as it was considered by our advisors to be jargon, and could result in restricted discussion to the constituent parts, such as ethics. A more conversational tone was adopted, addressing some of the companies' potential concerns; avoiding what might be seen as any moralising attitude or the pursuit of impractical theoretical outputs from academics; and included clear reference to links with established business organisations that were partners in the project.

With this new pitch greater interest in the project was generated, but no companies confirmed any commitment to engagement. It seemed there was still some confusion as to what the benefits of participating in the research were and what was required of the companies, especially in terms of the time commitment. Significant discussion at a project meeting came up with the idea of pitching the workshops as free 'innovation consulting' to see if that would impact the involvement of companies. In this rather lengthy pitch, it was possible to demonstrate knowledge of the issues cyber security companies faced.

Unfortunately, this new pitch did not work very well either, perhaps because of its length (six paragraphs and some bullet points), or perhaps because it seemed a bit too good to be true (in fact, one of the participant companies regularly checked to make sure they didn't have to pay for anything). Also, it seemed that the relatively lengthy time commitments envisaged ("less than a day and a half spread over a couple of months") were considered particularly onerous, and the collaborative working was seen as too complicated, in part due to the intellectual property that could be compromised if collaborative activities were undertaken. Further discussions within the project offered a revised and final (Appendix) research protocol, with two 2–2.5 h face-to-face workshops in which the researchers came to the companies. A revised sales pitch concentrated on the potential benefits for the companies from engaging in the activities, focusing on topics such as trust-building and ethics.

Another change in strategy was to become engaged in activities that the companies were running themselves. In this way, rather than asking companies to come into what they might perceive as an academic world somewhat detached from commerce; the academics would be working in the world of industry. This was complemented by engaging in talks and networking events (De Montfort University Cyber Forum, IOActive's HACK::SOHO, Malvern and South Wales Cyber Security clusters seminar sessions and workshops, a company launch) and speaking at industry venues. The ability to engage with the audience on the topics of ethics, responsibility and trust helped to validate the expertise of the researchers and the development of trust relationships with company representatives. For some companies, knowing that others had already taken up the offer also helped establish this trust relationship with project engagement and workshops taking place, sometimes reinforced through recommendations from their advisory boards.

Personal connections made through face-to-face discussions at networking events or talks also made a significant difference to the uptake of our subsequent workshops, compared with email introductions, and even more than cold-emailing. Frequently, the companies pointed to a specific set of issues they wished to be discussed in workshops, either problems they had encountered that we might give tailored advice on, or asking us to help them consider different options available to them as they moved from being a very small company of only a few employees to a more structured and larger company. Once again, the focus was around how the companies would benefit. They did not want to generally contribute to research without a well-thought-out set of benefits that they would receive in the process. Additionally (again reinforcing the importance of the interpersonal relationships) being able to show expertise in the specific area of cyber security (i.e. being able to 'talk shop') had a definite advantage in terms of showing trustworthiness and the relevance of the RI activities the companies were being asked to participate in.

Once the companies had taken up the offer, and the initial workshops were set up, some interesting issues around informed consent forms emerged. Discussing confidential business information is relatively taboo in cyber security as these companies are by nature generally quite secretive. It was necessary, therefore, to reinforce the initial trust that had been established through e.g. the use of appropriate consent forms, signing non-disclosure agreements, and other mechanisms. The informed consent procedures followed a fairly standard approach that is typical for university-led research – ensuring that participants understand what the research is about, what information will be taken, how the information can be used, and how they can withdraw from the study. For the workshops, the written work the participants developed and the discussions that were recorded (video or audio) were the main pieces of information taken from the experience.

Usually, for this sort of research, these procedures are easy to gain ethical approval for. This project was no different, and ethical approval was gained from the De Montfort University Ethics Review Board for the Faculty of Technology. However, the companies participating in the workshops, often with their legal advisors present, had difficulty with the (UK academic standard) consent documents. One company had issues with the representativeness of the discussion – with the employees in question being subject to non-disclosure agreements about company procedures and otherwise not speak for the company. Related issues were: How could they engage in this sort of research where they are being asked to discuss company approaches to responsible innovation? Were they speaking personally, or representing the company? After the CEO reassured the employees that they would not be breaking their contracts to discuss anything he or she was open to, the workshop continued. Another company asked that the researchers should also sign non-disclosure agreements about the specific company processes and procedures that might be discussed although all of these conditions were covered by the informed consent form and research ethics approval underpinning the research. Clearly the companies felt they needed an added level of security for their intellectual property. There was, furthermore, a seeming parallel between the lack of understanding of how university research projects function and the relationships between cyber security companies and end-users or clients (as previously discussed) who often don't understand how the cyber security technologies work.

With trust in the research process having been reaffirmed, the companies were prepared to trust the researchers with significant amounts of useful information to further understand the opportunities, challenges, costs and barriers to implementing RI practices in their businesses. This allowed unparalleled access to their processes and gave emphasis to the need for trust in the research process. Having succeeded in establishing such trust with four cyber security SMEs, a total of eight workshops took place.

6.4 Discussion

The lessons from the approaches discussed above are important in the context of recruiting and engaging with companies for academic research around RI. These may be generalisable and any recruitment strategy could adapt these lessons to their own specific industry sector. The lessons are illustrated in Table 6.1 and discussed below.

Table 6.1 Summary of findings

Coming down from the ivory tower	Standard ethical approaches may not be recognised	The need for expertise
Tailoring	Failure of standard academic approaches	One-to-one instead of one-to-many

6.4.1   The Importance of Coming Down from the Ivory Tower

One of the key lessons was that small companies in particular do not often have the resources to engage with research if it involves them coming to the researchers. More importantly, in going to the activities that the companies themselves initiated, a signal was sent that the researchers a) understood both their space, and that they had these activities in the first place; and b) were happy to engage on their terms (including accommodating and facilitating discussion on topics of particular concern to them). This helped to establish the element of trust whereby the companies would 'host' (food and refreshment and meeting venue) as well as engage with the researchers on a reciprocal basis.

6.4.2   Standard Ethical Approaches May Not Be Recognised

One of the more surprising lessons was the pointer to how much academic researchers may trust in ethical procedures and research ethics committee approvals granted for these sorts of activities. The fact that some of the companies required additional layers of protection for their intellectual property and procedural approaches was particularly interesting considering that they were, in fact, covered by the ethical approval processes. Is this a sign that there is little trust propensity for scientific research ethics processes outside of academia? Or is it more indicative of the particularly secretive natures of cyber security companies? No other sector companies engaged in our project had issues with the consent documentation, but perhaps this is because those other sectors addressed in the project (biomedicine and nanotechnology) are more closely aligned with traditional academic scientific research, where there is familiarity with and trust in these procedures.

It is important that this issue is considered by researchers when engaging with companies, and certainly those in the cyber security sector. It follows that the ability of companies to sign non-disclosure agreements that cover the same conditions as more standard academic consent procedures should be discussed with university legal services and ethics review committees, and legal teams within companies given time to investigate them. Additionally, fall-back options should be considered. For one company, for instance, workshops were only recorded audio, as video recording was considered too invasive.

6.4.3   The Need for Expertise in the Target Area

Throughout this whole procedure, the need for the researchers to 'prove themselves' as experts with reasonable knowledge in the specific sector area, and not just in applied ethics/responsible innovation was clearly important. A significant understanding of technical issues was definitely advantageous when working with the companies. Being able to tailor questions to help each company delve into the ethical questions surrounding their specific lines of work was very helpful to get detailed, in depth, responses. Cyber security is a widely varied sector, and with expertise of many of the different areas it is clearly easier for the researcher to establish trustworthiness, and more likely that the company will have a trust propensity for the researchers. Indeed, the company's understanding must be that the researchers will understand some of the complexities of the sector and their business and, therefore, be able to use the research outcomes effectively.

Similarly, the use of "known experts" as part of the pitch, particularly those in cyber security companies' areas of interest, including the in-house expertise of cyber security researchers at the university, the local police, business support organisations, and others, improved the credentials of the research team, showing that we were engaged with other organisations and businesses outside of the university.

6.4.4 Tailoring Is Advantageous

Expertise in the subject area can also help to fulfil another requirement, that of tailoring the discussions to the specific company. The cyber security workshops were characterised by co-creation activity by peers and were conducted with several members of the same company. This allowed for tailoring of the information provided to the company, rather than a more generic approach. Such tailoring requires more understanding of the company involved, and expertise on the part of the researchers to be able to analyse and report back on the results. By following this approach, the results from the cyber security workshops allowed a richer set of outcomes than those which arose from the 'collective' approach to workshops that were undertaken for the biomedicine and nanotechnology sectors elsewhere in the project.

6.4.5 The Failure of Standard Academic Approaches

Standard academic approaches for research recruitment generally include calls for participation via email lists, or newsletters, or other methods that are often picked up by multipliers. These kinds of 'passive consumption' requests for engagement were largely unsuccessful in this study. Unlike with academic calls for papers or similar, these kinds of activities are not part of the day-to-day business of cyber security companies, which may explain why such calls were regularly ignored. Other standard academic approaches to potential participants, such as offering to pay for travel and accommodation, food, etc., also did not work. This may be explained by the fact that many of the SMEs engaged with were time-poor, with several potential participants dropping-out of the process due to lack of time or the inability to agree a mutually convenient time. Clearly to contribute a day or two of their time is overly burdensome for many SMEs, even with financial compensation. The fact that the researchers were willing to travel to the companies was well-received by the companies involved, as was the reduction of the time investment required.

6.4.6   One-to-One Instead of One-to-Many

Finally, the advantages of sending personalised, follow-up emails after a personal introduction or meeting at a networking or talk event are significant. As has been noted, the original approaches of sending information to potentially interested parties via multipliers (e.g. the university's cyber security network, the UK cyber security forum, and larger multipliers such as more general business networks) were largely unsuccessful. Large-scale advertising allows for relative anonymity and, it is suggested, can lead to a lack of response. Ignoring *personal* emails after initial connections are made is much less socially acceptable and, even when invitations are declined, these refusals can offer useful insights into the reasons (e.g. time constraints, concerns about confidentiality). Additionally, recommendations from boards of trustees/advisory boards for their companies to participate, as well as their having knowledge that other well-respected companies are participating, helps increase the predisposition to take part. The trust companies have in advice from these boards also contributes to the overall trust propensity of the cyber security practitioners in the researchers themselves.

6.5   Conclusion and Recommendations

These experiences describe how hard it sometimes is to recruit companies to work with RI research projects. Often there are conflicting ideas of roles, benefits, what is required, and what outputs are created. In moving from an academic sphere to a business sphere, going into their world and becoming involved in their events, approaches, and ultimately understanding their positions, it was possible to recruit companies who not only initially engaged, but over time became longer-term partners with the project, offering to go above and beyond the minimal engagement requirements. These interactions point to a high level of trust between the researchers and the companies: not just that the researchers were trusted, but that they were trustworthy. This reflects, it is considered, the desire that the companies have to, themselves, be seen as trustworthy beyond the cyber security sector: pointing to such "trust" being a key reason for engaging with the activities during the workshops. This validates the usefulness of locating a key value that the industry is likely to engage with, and that aligns with RI principles and practices, in order to use it as a method for engagement.

Overall, the following approaches worked best for engaging with cyber security companies about RI:

- Removing the academic terminology of "responsible research and innovation" as a concept in its own right, and talking about its constituent parts using industry language. Hence, the use of the simpler term "responsible innovation", acknowledging the overlap with the more familiar concept of corporate social responsibility.
- Being clear about the benefits of ethical approaches in commerce.
- Making a positive effort to understand the commercial context within which SMEs operate (i.e. through engaging in or speaking at their events), rather than expecting them to come into the academic world.
- Engaging with external advisory organisations to boost credentials and trustworthiness.
- Extending academic knowledge around responsible innovation and ethics in order to understand the key technical and commercial dilemmas, challenges and opportunities that confront companies in the sector in question.
- Minimising the requirements for companies to participate (e.g. time, travel, etc.).
- Being positioned to assist with any particular ethical dilemmas or issues faced by the companies.
- Engaging in personalised and often face-to-face discussions with key members of the companies in order to demonstrate understanding, and to establish a rapport conducive to outcomes within workshops.

Some of these approaches may be more specific to cyber security companies, but there are wider lessons for other sectors. Perhaps most notable (and generalisable) is the importance of understanding the sector in question and its commercial context in order to engage with the staff, often at a senior level, of SMEs. This positions the researcher more clearly as an equal in the search for insights and truths that the workshops can reveal. Linked with this is the need not to offer RI as a model or blueprint, but rather to demonstrate knowledge of the sector; personalise and tailor information to the specific company; and to focus on those components of RI which are already recognised by the company.

Image Credits Tower by iconcheese from the Noun Project
contract by Templet from the Noun Project
consulting by Vectors Market from the Noun Project
Tailor by Pham Duy Phuong Hung from the Noun Project
Recruitment by Massupa Kaewgahya from the Noun Project
Conversation by Olivia from the Noun Project

Appendix

Dear _____,

As a security company, you're probably very concerned about ethics, and ensuring your business acts as responsibly as possible. What we want to do is to help your company be even more ethical in your business practices.

We want to be pragmatic, useful, and responsive to your company's needs and goals.

Much like the security sector sells an idea – that security needs to be built in from the beginning – we will convince you that if you build in responsible and ethical practice from the beginning, you'll benefit from it in the medium-long term through:

- better relationships with clients;
- broader and more sensitive outreach and sales approaches;
- higher levels of client trust in your company;
- a more embedded community presence;
- and an agility for future challenges and opportunities.

We will work directly and confidentially with you and your company, identifying your areas of good practice and injecting good practice identified by interviews with practitioners, CEOs, and developers of other tech companies. We have successfully done this with the health technology sector in the past, and now we want to open up our methods to the security sector.

We want your company to be prepared for what the future might bring – 2, 5, even 10 years down the line, and help you to put good practice in place to be able to deal with these challenges and opportunities. You'll also learn how to use our techniques to help potential clients think about their own futures – and how security can benefit them.

We'll need around 5 h of your time total, spread over 2 face-to-face meetings where we come to you, and a couple of short follow-up phone calls/emails after each meeting. In between, we will integrate expert opinion from our research for the COMPASS project, the East Midlands Police, academic security researchers, business support organisations such as B Labs and EBN Innovation Network, and professional organisations to help you look above and beyond your everyday practice.

You'll get a tailored, future-looking roadmap to practically implement responsible and ethical practice in your company, so you can benefit from being more trustworthy, learn from our methods, and end up with a more agile, future-looking company that can be relied on by customers and the public to behave ethically and responsibly.

For more information please contact …

Sincerely,

References

Department for Business, Energy & Industrial Strategy. (2017). UK innovation survey 2017: headline findings [WWW Document]. GOV.UK. https://www.gov.uk/government/statistics/uk-innovation-survey-2017-headline-findings. Accessed 11.8.18.

European Commission. (2011). A renewed EU strategy 2011–14 for corporate social responsibility', Communication from the commission to the European Parliament, the council, the European economic and social committee and the Committee of the Regions COM/2011/0681 final.

European Commission, Directorate-General for Research and Innovation. (2013). *Options for strengthening responsible research and innovation*. Luxembourg: EUR-OP.

Hughes, M. (2015). *Full or responsible disclosure: How security vulnerabilities are disclosed* [WWW document]. MakeUseOf. http://www.makeuseof.com/tag/responsible-disclosure-security-vulnerabilities/. Accessed 13.12.17.

Lefkowitz, J. (2017). *Responsible disclosure – Critical for security*. Critical for Intelligence | SecurityWeek.Com [WWW Document]. URL http://www.securityweek.com/responsible-disclosure-critical-security-critical-intelligence. Accessed 13.12.17.

Owen, R., Stilgoe, J., Macnaghten, P., Gorman, M., Fisher, E., & Guston, D. (2013). A framework for responsible innovation. In R. Owen, J. Bessant, & M. Heintz (Eds.), *Responsible innovation* (pp. 27–50). Chichester: Wiley. https://doi.org/10.1002/9781118551424.ch2.

Stahl, B., Flick, C., Mantovani, E., Borsella, E., Porcari, A., Barnett, S. J., Yaghil, A., Ladikas, M., Hahn, J., Obach, M., Garzo, A., Schroeder, D., Chatfield, K., Antoniou, J., Paspallis, N., Brem, A., Yaghmaei, E., Brey, P., Søraker, J. H., Gauttier, S., Gurzawska, A., Ikonen, V., Leikas, J., & Mäkinen, M. (2017). Benefits of responsible research and innovation in ICT for an ageing society. *Responsible-Industry Project*. https://doi.org/10.5281/zenodo.1050357.

Von Schomberg, R. (2013). A vision of responsible research and innovation. In R. Owen, M. Heintz, & J. Bessant (Eds.), *Responsible innovation. Managing the responsible emergence of science and innovation in society* (pp. 51–74). Wiley.

Chapter 7
Towards a Business Case for Responsible Innovation

Norma Schönherr, André Martinuzzi, and Katharina Jarmai

Abstract There is still work to be done in conceptualizing how responsible innovation applies to business. Lessons can be drawn from adjacent fields of inquiry such as sustainability-oriented or social innovation. However, the central challenge of developing a business case for responsible innovation requires additional insights into how responsible innovation may support companies in generating competitive advantage, and what levers can be effectively employed to engage business. This final chapter summarises the most important lessons learned from the contributions to this volume. Based on these insights, the authors develop the outlines of a business case for responsible innovation. In doing so, they show that responsibility and innovation can mutually strengthen each other. Such a synergy between responsibility and innovation may help to maintain trust in business' ability to drive desirable social change while improving innovation performance.

Keywords Responsible innovation · Business case · Corporate responsibility · Competitive advantage

7.1 Introduction

In the broadest sense, the purpose of a business case is to make relevant decision-makers aware of a new business opportunity, educate them as to how an organization can seize this opportunity, and justify the costs and potential risks of taking action against the benefits to be expected. A compelling business case also needs to present a range of options, with reasons for rejecting or carrying forward each proposed option. As such, a business case is necessarily also a platform for deliberation on the merits of both the business opportunity presented and the options

N. Schönherr (✉) · A. Martinuzzi · K. Jarmai
Institute for Managing Sustainability, WU Vienna University of Economics and Business,
Vienna, Austria
e-mail: norma.schoenherr@wu.ac.at; andre.martinuzzi@wu.ac.at; katharina.jarmai@wu.ac.at

© The Author(s) 2020   85
K. Jarmai (ed.), *Responsible Innovation*, SpringerBriefs in Research and
Innovation Governance, https://doi.org/10.1007/978-94-024-1720-3_7

proposed for seizing the opportunity. This chapter proposes this kind of business case for responsible innovation (RI) as food for thought to practitioners and academics alike. As such, this chapter strives to provide a framework for readers to reflect on how the concept of responsible innovation applies to their day-to-day practice.

We first summarise key lessons learned from the chapters in this volume and develop the foundations of RI in a business context. Building on these insights, we elaborate on RI as a business opportunity. We then propose six distinct but mutually supportive pathways for leveraging RI to ensure a social license to operate, maintain consumer trust, secure competitive advantage, enhance innovation performance, and build capacity within organizations. We conclude by highlighting open questions and presenting a glimpse of the road ahead towards a business case for responsible innovation.

7.2 Foundations of a Business Case for Responsible Innovation

The chapters in this volume illustrate that RI is emerging as a new field in the continuing discourse on the role and responsibility of business in society (Martinuzzi et al. 2018). It has the potential to advance this discourse in light of two key competitive factors: innovativeness in the context of an increasingly intensive race for the "next big thing", and trust (of customers, employees, investors and other stakeholders) in business. The first relates to the accelerating race to innovate in order to stay competitive in a rapidly changing world (Stata 1994; Schwab op. 2016). The second concerns the need to maintain public trust through innovations that generate social value in addition to economic returns (Pirson et al. 2017; Lewicki et al. 1998). Both aspects are equally important in developing a business case for RI.

The concept of RI is embedded in an ongoing debate on the broader responsibility of business towards society and the environment (Bansal and Song 2017; Carroll 2015). Concurrently, it relates corporate responsibility (CR) to one of the core functions of many companies: innovation is for many a key requirement to stay competitive in light of ongoing digitalization, globalization and rapidly changing markets (Crossan and Apaydin 2010; Mone et al. 1998; Dess and Picken 2000). Eight of the ten most valuable publicly traded firms in the world in 2018 were technology companies, with a combined market value of over US$5trn (Forbes 2018). At the same time, especially innovation-intensive and technology companies face increasing expectations that they will contribute to coping with the technological, social and political impacts generated by their innovations. For instance, Youtube, Facebook and other platforms are part of a controversial and continuous social debate as to whether and to what extent they should assume responsibility for the contents their users publish on their platforms (The Economist 2018; The Guardian 2017).

Concurrently, Google is under continuous pressure to demonstrate what their company credo, "Don't be evil", means in practice, and whether it is socially and ethically acceptable to provide an adapted search engine for the Chinese market that leverages their technological know-how but may enable illiberal governments to monitor what information their citizens can access (Bloomberg Businessweek 2018; Fortune Magazine 2018).

The current speed of innovation goes hand in hand with a general drop in trust in societal institutions such as governments and media, but also companies (Pirson et al. 2019). This leads many people to be wary of new technologies. With the fall of trust, many now lack full belief that the overall system is working for them. In this climate, people's societal and economic concerns, including globalization, the pace of innovation and eroding social values, turn into fears, spurring the rise of populist actions, on the one hand, and anti-business sentiment, on the other hand (Gardels and Berggruen 2017). For instance, the Edelman Trust Barometer (2017), an annual survey of more than 33,000 respondents across 28 countries, revealed that 51% of respondents were concerned about the pace of innovation and 22% expressed fear that technological innovations were happening too quickly and leading to changes that were not good for them. In this context, about two thirds of respondents did not believe information shared by the CEOs of companies was credible, and expected business to lead through action rather than words. In this vein, 75% of respondents agreed that companies can and should take specific actions that both increase profits and improve the economic and social conditions in the community where they operate (Edelman 2017). That is, companies are expected to create shared value (Porter and Kramer 2011) and take on responsibility beyond the boundaries of their organization that is commensurate with the power they wield over consumers' lives. In light of this "techlash" (a combination term used to designate a societal backlash against technology), companies are increasingly called upon to take measures to ensure that the benefits of innovation are not overtaken by detrimental social and environmental impacts (Voegtlin and Scherer 2017). While the chapters in this volume have shown that there is still considerable debate about the exact nature of such measures, there is a consensus that they should refer both to the innovation process (how companies innovate), and its marketable results (products, services and business model innovations) (Lubberink et al. 2017; Stilgoe et al. 2013).

The issues raised above are not mundane questions. What is more, these questions also apply to Small and Medium Sized Enterprises (SMEs), especially in highly innovative sectors (Halme and Korpela 2014; Auer and Jarmai 2018). The cases of Yoti and AppNps, presented in Chaps. 2 and 6 of this volume, aptly show that innovations originating in SMEs also engender new concerns, e.g. in relation to data privacy, or the potential long-term toxicity of new materials. These cases also illustrate how SMEs that leverage new technologies to provide new services can cope with these challenges to ensure that their innovations ultimately improve people's lives.

Companies need not start from scratch on the journey towards embedding RI (van de Poel et al. 2017). The discourse on responsible business is mature, and many instruments are already available to companies for discharging their responsibility towards society (Iatridis and Schroeder 2016). RI can draw on more than three decades of experience with practices such as sustainability reporting (Hahn and Kühnen 2013), technology assessment (Grunwald 2014), human-centered design (Buchanan 2001), open innovation (Bogers and West 2012) and many others. Chaps. 3 and 4 of this volume show how RI can leverage the thinking and instruments developed in adjacent areas, notably sustainability-oriented innovation and social innovation (also see Lubberink et al. 2017). Chap. 3 shows how the focus on designing for sustainability and anticipating impacts can inform RI in SMEs, and Chap. 4 highlights the potential of collaborative interactions between innovative companies and communities. Both chapters present cases where businesses have reconnected with the communities they serve, while becoming more prosperous and successful. Often this leads to both immediate benefits for the business but also sets the framework for a long-term strategy that could potentially initiate and support both social and environmental change (Goodman et al. 2017; Gurzawska et al. 2017).

The fundamentally value-based nature of responsible innovation has been a tenet throughout the book. RI is by definition a normative, values- and purpose-driven concept, which requires the alignment of economic, societal and environmental business goals. As concluded in Chap. 6 of this volume, innovative SMEs have to decide for themselves what they consider as their responsibility towards society, and how much this decision is based on success measured in economic terms. As such, the trail-blazers in the realm of RI (some of them have been presented in this volume) are united in that they acknowledge an intrinsic motivation to engage with RI, going beyond short-term profit generation. Other drivers for engaging with RI may derive from external pressure exerted upon companies (e.g. legal pressure or funding and financing requirements), or mediated through direct stakeholder relationships that firms maintain (e.g. with peers, communities, or consumers). The lesson we learn from this is that the exact configuration of a business case will depend on determining the right fit between the drivers of responsible innovation (see Table 7.1) within the company, the measures it is willing to take, and the authenticity and effectiveness of both in light of the external environment and the relationships that companies engage in (Schaltegger and Wagner 2011; Goodman et al. 2017).

Finally, the individual contributions in this volume have shown the breadth and diversity of the discussions around RI in academia, but also the policy and business sectors. The cases presented in the individual chapters show that implementing RI in day-to-day practice is challenging and requires the continuous identification, combination, and review of instruments available to business for discharging their responsibility (Iatridis and Schroeder 2016; Fisher and Rip 2013). A broader business case will have to acknowledge these challenges. Responsible innovation engenders both costs and benefits. Striving for balance between the two ultimately determines the business case for responsible innovation.

Table 7.1 Drivers of responsible innovation derived from the chapters of this volume

	Driver	Description
Internal drivers (= moral standards & economic motivations)	Intrinsic motivation of key individuals	Moral standards of high-level decision-makers within firms, which evolve around the morality of products and services, their effects on human beings and social issues within global value chains
	Economic motivation of key individuals	Perceived instrumental value of RI for generating added value for the company through a range of aspects such as risk reduction, cost efficiency, reputational effects, market differentiation or market development
External drivers (= external context factors)	Cultural setting	Ethical and belief systems prevalent in specific regions, communities and/or countries
	Legal frameworks	Regulations pertaining to innovation processes and outcomes (particularly in highly regulated industries, such as healthcare)
	Funding and financing requirements	Inclusion of RI criteria in relevant public funding programmes or as a basis for obtaining finance
Relational drivers (= stakeholder relations & public relations)	Social license to operate	Maintenance of the ongoing acceptance of companies, their innovations and practices by its stakeholders as well as the general public
	Best practice	Implementation of generally accepted operating procedures for innovation management in a given sectoral or industry context
	Reputation	Maintenance of a generally positive public perception of the financial, social and environmental impacts attributed to the company over time

7.3   Leveraging Responsible Innovation to Create Business Opportunities

Companies already successfully engage with many practices that fall under this umbrella term. However, these practices are frequently disjointed, distributed across business functions or unconnected to the core of innovation within companies (Lubberink et al. 2017). RI is an opportunity for companies to integrate these elements and practices in a coherent framework for better management of innovation processes and better results, for the mutual benefit of companies, society and the environment (Stilgoe et al. 2013). The potential business benefits of engaging with RI are many, and include enhanced trust, creativity, openness to new business opportunities beyond the boundaries of the company, and improved capacity for engaging with peers, consumers and communities, as well as reduction of risk and uncertainty in increasingly fast-paced innovation cycles. Chaps. 5 and 6 of this volume illustrate concrete cases of how RI can help companies realize these benefits and thereby turn responsible innovation from 'a drain on company resources' into 'competitive advantage'.

Drawing on these specific cases, we outline six pathways that companies may leverage to pursue business opportunities in the spirit of RI. This includes capacity building as prerequisite for leveraging RI, respect for ethical limitations, an inclusive approach to innovation, the careful balancing of interests, anticipation of (potential) impacts, as well as broadening one's perspective to consider systems dynamics which affect and are affected by the company.

7.3.1 Capacity Building for Responsible Innovation

Responsible innovation requires building up new skills and capacities within companies. In many cases, it can also mean linking and combining the capacity already present within different parts of a company. Especially at the beginning of their responsible innovation journey, managers need to be clear as to the purpose and motivation behind engaging with RI practice in order to limit complexity and clearly identify learning opportunities. In this vein, it is useful to consider whether companies wish to leverage RI to review and improve their internal processes towards achieving their goals, i.e. internal orientation, or whether they wish to respond to expectations held by stakeholders and the general public, i.e. external orientation (see Fig. 7.1).

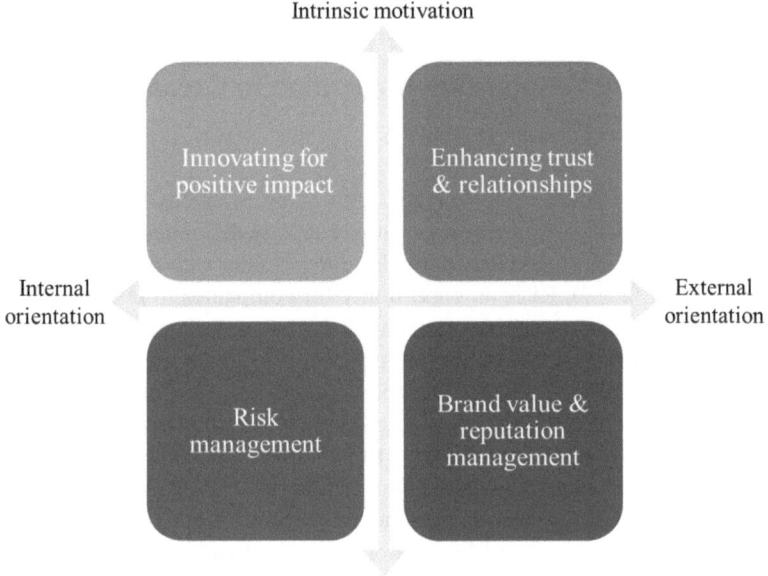

Fig. 7.1 Business objectives related to RI in light of motivation and orientation of the organisation

Where the motivation to engage with RI derives from intrinsic moral convictions, the achievement of company objectives is tied to optimizing innovation processes to generate solutions that maximize positive societal impacts while minimizing detrimental effects. This is frequently the case for social enterprises pursuing a specific social purpose (see Chap. 3 of this volume). For high-risk or socially contested areas of innovation, RI may be all about enhancing trust and relationships between the company and its stakeholders. For companies with a primarily economic motivation, business opportunities arise from leveraging RI for risk management, and brand value & reputation management. Defining a clear purpose enables companies to pilot RI approaches and learn from the experience.

In addition, companies wishing to engage with RI may want to assess their current practices and policies against RI principles to identify strengths and areas of improvement. Building on the (sometimes tacit) knowledge already present in companies presents a chance to leverage RI for organizational learning. To support such self-assessment processes, the EU-funded COMPASS project has developed a comprehensive self-check tool[1] translating the concept of responsible innovation into concrete corporate practices and policies. This allows for an adaptable judgement of company strengths and weaknesses with regard to RI. Lessons learned from applying the tool are complemented with incentives for continuous improvement, such as positive scoring and alignment to a road mapping method for developing strategy in line with RI.

7.3.2 *Respecting Ethical Constraints*

Many companies feel the pressure to innovate ever faster to stay competitive. For citizens, the speed of development of new products, services, technologies and business models can be overwhelming. Policy makers struggle to provide the frameworks and rules that can maximise the potential of innovation for the common good while effectively dealing with the risks and the ethical concerns they raise. This creates grey areas and uncertainty (Stern 2017).

Ethical frameworks can help deal with this problem by outlining the values, concerns and limitations that research, development and innovation should respect. Ethical constraints can vary in different cultural and legal contexts – ignoring ethical constraints, however, poses the risk of losing social license to operate. This is why, maybe counterintuitively, ethical constraints are good news for innovation. They provide guidance in a space of uncertainty and ensure that new technologies are not only acceptable, but also desirable for society.

Some companies profit from pre-existing sectoral and industry ethical guidelines. However, companies in sectors where such guidelines are not yet established or are insufficient for their purposes need not give up. As the example of the nanotechnology company AppNps (see Chap. 5) illustrates, investing in the creation

---

[1] https://innovation-compass.eu/selfchecktool

of ethics guidelines that go above and beyond legal compliance can support employee retention and prevent resistance to innovations in socially contested areas.

7.3.3 Taking an Inclusive Approach

Almost all creativity that goes into innovation is geared toward solving problems. Many companies have already learned that diverse teams and diversity in management can be great drivers of creativity, which are better at solving problems than homogeneous ones. When it comes to innovating for users and application areas outside the company, involving those affected can be equally effective in enhancing creative thinking and problem solving. This is illustrated by the example of the smart meter company ambiact (see Chap. 5). The case specifically points to cost savings that can be realized through early user involvement, by speeding up the overall process from idea to marketable product.

Concepts like social innovation, open innovation and lead-user innovation have provided methods for leveraging this potential for companies. What RI adds to the equation is the ambition to make involvement inclusive, meaningful and beneficial, not only for the innovators, but also for the diverse stakeholders involved in open and lead-user processes (Bessant 2013). Considering both sides as equally important and striving for true co-creation can help overcome innovation barriers and increase societal acceptance, desirability and accessibility of innovation outcomes. In addition, the inclusion of societal actors outside the immediate target group of innovations may point to completely new application areas for existing technologies or enable the identification of new customer groups that were not previously reached (Heeks et al. 2014).

7.3.4 Balancing Interests

A proactive approach to balancing interests is critical for realizing business opportunities from RI for three reasons. First, different groups and individuals may have very different expectations toward innovations and those that develop them. While some may hope to directly benefit and see an improvement of some sort in their life, others may worry about unintended consequences (such as the potential for weaponization of new technologies or potentially harmful long-term effects to health). Second, many societal challenges, such as climate change or social equity, are contested issues, which limits the ability of stakeholders to find common ground when defining what is responsible. When consensus is lacking, ambiguities arise and innovations are likely to encounter a 'techlash' rather than societal acceptance. Third, openness and inclusiveness in innovation processes pose new challenges in light of the prevalent conception of innovation, which holds that innovations are rooted in information asymmetries in the market. Companies may therefore have marked incentives not to engage with stakeholders in sensitive innovation processes. All of these concerns are valid and need to be proactively addressed.

Chap. 6 of this volume illustrates how continuous engagement and dialogue between researchers and cybersecurity companies over an extended period was an important precondition for establishing trust, acceptance, and openness around innovation processes. RI encourages innovators to take both company and stakeholder needs and concerns seriously. This means communicating with partners and stakeholders on equal terms – for instance by explaining technologies-in-use rather than in abstract technical terms. It also means being transparent and accountable about how innovations are created, implemented and scaled. This creates trust and limits the risk of rejection of innovations at a late stage of development or market deployment.

7.3.5 Anticipating Impacts

Technology impact assessment is a well-established practice in many companies. However, such assessments are often one-time exercises at a relatively advanced stage of innovation processes, when significant costs have already been incurred. Systematically embedding impact assessment into all stages of the innovation process can help to recognize risks and potentially detrimental impacts at an early stage (Grunwald 2014). The case of carpet manufacturer Mission Zero (see Chap. 5) aptly illustrates how companies can be economically successful while designing for low (or even zero) negative impact.

However, anticipating impacts is not just about avoiding harm but also about actively seeking business opportunities in areas where innovations can do the most good. Involving users and other stakeholders in evaluating potential risks and impacts beyond purely technical concerns is an important part of this (van den Hoven 2013). Failing sooner and earlier in the innovation process can help avoid sunk costs and redirect innovation processes toward those fields where innovators can make a positive contribution to society (Doorn 2013). This idea is at the heart of the following pathway, co-designing systems.

7.3.6 Co-designing Systems

Asking the question "Where can the key competences of my company create the most positive impact?" can be a source of inspiration and new business opportunities. Some of the most radical and successful innovations have come from re-designing whole systems, rather than from improving existing products or technologies. The sharing economy, for instance, does not rely on new innovative products but has redefined the way products and services are distributed, consumed, and paid for.

Co-designing innovations at the systems level helps to find holistic solutions for complex problems. Looking beyond the boundaries of a company's core business also opens up new spaces for innovation and can help identify new business opportunities. For instance, the Business and Sustainable Development Commission, a high-level forum of leaders from business as well as other private sector and civil society organizations, has examined economic systems with high potential for transformative innovations with significant societal impact. The report identifies 60 market opportunities related to food and agriculture, cities, energy and materials, as well as health and well-being. The business case for companies to address these areas is strong: at least US$12 trillion in new business opportunities are expected (Business and Sustainable Development Commission 2017).

Co-designing systems may present the most disruptive approach for a company when following RI, and the co-design of systems is the most challenging pathway included in this chapter, as it requires knowledge of global trends and system dynamics (Stata 1994; Herrera 2015; Business and Sustainable Development Commission 2017). In addition, it requires managers to engage with a much more complex system of relationships and responsibilities, many of them not under the exclusive control of any one company (Voegtlin and Scherer 2017; Schönherr et al. 2017).

7.4   Open Questions and the Road Ahead

This chapter has provided a synthesis of the lessons learned from the individual contributions in this volume, and has developed the outlines of a business case for responsible innovation. However, open questions remain.

There is work to be done to fully appreciate the potential tension between the ethical, social, and environmental mandate of RI and the profit-oriented rationale of micro-economic decision-making. For instance, Chap. 2 of this volume shows that the job of translating the principles of RI into business-relevant language and concrete managerial practice is far from accomplished. Chaps. 5 and 6 allude to potential returns of implementing RI for competitive advantage while also acknowledging that win-win situations are not guaranteed, or even likely to arise in all areas. There are cases where a business case for responsible innovation may not materialise because of marked conflicts between the adoption of RI and commercial interests. The direct and indirect impacts of innovation are difficult to quantify, and economic returns frequently depend on the behaviour of external stakeholders, such as customers, peers, and regulatory bodies. In addition, innovations may become transformative game changers, which may entail societal effects that cannot be foreseen with any certainty.

While the contributions in this volume provide an overview of the breadth of the discourse around RI in a business context, there are significant gaps in what we know about what is required to build a comprehensive business case for responsible innovation. Future research will need to expend some effort on clarifying the importance of factors such as industry sector, firm size, organizational culture, governance structure, regulatory framework, and others that have been shown to be relevant for embedding responsibility into industry. An examination of such factors in relation to drivers of innovation (see Table 7.1) will require particular attention. A key aspect that needs to be considered when discussing the business case for RI refers to the context that sets incentives and boundaries for company action, not only including regulation and legislation, but also customs and culture, which can shape the way RI is perceived and implemented.

Open questions also remain in light of the shared responsibility between a multitude of actors involved in innovation. Future work might, for instance, examine the interfaces and value chains where industry and societal groups jointly negotiate the meaning of responsibility. The opportunities related to a more networked understanding of RI that goes beyond the focus on individual companies dominating the discourse may help to fully appreciate the potential of a more collaborative approach to RI, and provide new avenues for eliciting how both the costs and benefits can be shared across the actors involved in innovation processes and outcomes as the emerging discussion on a business case for responsible innovation evolves.

References

Auer, A., & Jarmai, K. (2018). Implementing responsible research and innovation practices in SMEs: insights into drivers and barriers from the Austrian medical device sector. *Sustainability, 10*, 17. https://doi.org/10.3390/su10010017.

Bansal, P., & Song, H.-C. (2017). Similar but not the same: differentiating corporate sustainability from corporate responsibility. *Academy of Management Annals, 11*, 105–149. https://doi.org/10.5465/annals.2015.0095.

Bessant, J. (2013). Innovation in the twenty-first century. In R. Owen, J. R. Bessant, & M. Heintz (Eds.), *Responsible innovation: Managing the responsible emergence of science and innovation in society* (Vol. 28, pp. 1–25). Chichester/West Sussex: Wiley.

Bloomberg Businessweek. (2018). *Google in China: When 'Don't Be Evil' met the great firewall.* https://www.bloomberg.com/news/features/2018-11-08/google-never-stopped-trying-to-go-to-china. Accessed 1 Mar 2019.

Bogers, M., & West, J. (2012). Managing distributed innovation: Strategic utilization of open and user innovation. *Creativity and Innovation Management, 21*, 61–75. https://doi.org/10.1111/j.1467-8691.2011.00622.x.

Buchanan, R. (2001). Human dignity and human rights: Thoughts on the principles of human-centered design. *Design Issues, 17*, 35–39. https://doi.org/10.1162/074793601750357178.

Business and Sustainable Development Commission. (2017). *Better business, better world.* http://report.businesscommission.org/.

Carroll, A. B. (2015). Corporate social responsibility. *Organizational Dynamics, 44*, 87–96. https://doi.org/10.1016/j.orgdyn.2015.02.002.

Crossan, M. M., & Apaydin, M. (2010). A multi-dimensional framework of organizational innovation: A systematic review of the literature. *Journal of Management Studies, 47*, 1154–1191. https://doi.org/10.1111/j.1467-6486.2009.00880.x.

Dess, G. G., & Picken, J. C. (2000). Changing roles: Leadership in the 21st century. *Organizational Dynamics, 28*, 18–34. https://doi.org/10.1016/S0090-2616(00)88447-8.

Doorn, N. (Ed.). (2013). *Early engagement and new technologies: Opening up the laboratory* (Philosophy of engineering and technology) (Vol. 16). Dordrecht: Springer.

Edelman. (2017). *2017 Edelman TRUST BAROMETER™- Global results*. https://www.slideshare. net/EdelmanInsights/2017-edelman-trust-barometer-global-results-71035413

Fisher, E., & Rip, A. (2013). Responsible innovation: Multi-level dynamics and soft intervention practices. In R. Owen (Ed.), *Responsible innovation: Managing the responsible emergence of science and innovation in society* (pp. 165–183). Chichester: Wiley.

Forbes. (2018). *The 100 largest companies in the world by market value in 2018 (in billion U.S. dollars). Statista – The Statistics Portal*. https://www.statista.com/statistics/263264/top-companies-in-the-world-by-market-value/. Accessed 1 Mar 2019.

Fortune Magazine. (2018). *Google employees revolt against censored search engine for China*. http://fortune.com/2018/11/27/google-china-project-dragonfly-employee-objections/. Accessed 1 Mar 2019.

Gardels, N., & Berggruen, N. (2017). Salvaging globalization. *New Perspectives Quarterly, 34*, 67–79. https://doi.org/10.1111/npqu.12070.

Goodman, J., Korsunova, A., & Halme, M. (2017). Our collaborative future: Activities and roles of stakeholders in sustainability-oriented innovation. *Business Strategy and the Environment, 26*, 731–753. https://doi.org/10.1002/bse.1941.

Grunwald, A. (2014). Technology assessment for responsible innovation. In J. van den Hoven (Ed.), *Responsible innovation: Innovative solutions for global issues* (pp. 15–31). Dordrecht: Springer.

Gurzawska, A., Mäkinen, M., & Brey, P. (2017). Implementation of Responsible Research and Innovation (RRI) practices in industry: Providing the right incentives. *Sustainability, 9*, 1759. https://doi.org/10.3390/su9101759.

Hahn, R., & Kühnen, M. (2013). Determinants of sustainability reporting: A review of results, trends, theory, and opportunities in an expanding field of research. *Journal of Cleaner Production, 59*, 5–21. https://doi.org/10.1016/j.jclepro.2013.07.005.

Halme, M., & Korpela, M. (2014). Responsible innovation toward sustainable development in small and medium-sized enterprises: A resource perspective. *Business Strategy and the Environment, 23*, 547–566. https://doi.org/10.1002/bse.1801.

Heeks, R., Foster, C., & Nugroho, Y. (2014). New models of inclusive innovation for development. *Innovation and Development, 4*, 175–185. https://doi.org/10.1080/2157930X.2014.928982.

Herrera, M. E. B. (2015). Creating competitive advantage by institutionalizing corporate social innovation. *Journal of Business Research, 68*, 1468–1474. https://doi.org/10.1016/j.jbusres.2015.01.036.

Iatridis, K., & Schroeder, D. (2016). *Responsible research and innovation in industry*. Cham: Springer.

Lewicki, R. J., McAllister, D. J., & Bies, R. J. (1998). Trust and distrust: New relationships and realities. *Academy of Management Review, 23*, 438–458. https://doi.org/10.5465/amr.1998.926620.

Lubberink, R., Blok, V., van Ophem, J., & Omta, O. (2017). Lessons for responsible innovation in the business context: A systematic literature review of responsible, social and sustainable innovation practices. *Sustainability, 9*, 721. https://doi.org/10.3390/su9050721.

Martinuzzi, A., Blok, V., Brem, A., Stahl, B., & Schönherr, N. (2018). Responsible research and innovation in industry—Challenges, insights and perspectives. *Sustainability, 10*, 702. https://doi.org/10.3390/su10030702.

Mone, M. A., McKinley, W., & Barker, V. L. (1998). Organizational decline and innovation: A contingency framework. *Academy of Management Review, 23*, 115–132. https://doi.org/10.5465/amr.1998.192965.

Pirson, M., Martin, K., & Parmar, B. (2017). Formation of stakeholder trust in business and the role of personal values. *Journal of Business Ethics, 145*, 1–20. https://doi.org/10.1007/s10551-015-2839-2.

Pirson, M., Martin, K., & Parmar, B. (2019). Public trust in business and its determinants. *Business & Society, 58*, 132–166. https://doi.org/10.1177/0007650316647950.

Porter, M. E., & Kramer, M. R. (2011). The big idea: Creating shared value. *Harvard Business Review, 89*, 1.

Schaltegger, S., & Wagner, M. (2011). Sustainable entrepreneurship and sustainability innovation: Categories and interactions. *Business Strategy and the Environment, 20*, 222–237. https://doi.org/10.1002/bse.682.

Schönherr, N., Findler, F., & Martinuzzi, A. (2017). Exploring the interface of CSR and the sustainable development goals. *Transnational Corporations, 24*, 33–47. https://doi.org/10.18356/cfb5b8b6-en.

Schwab, K. (Ed.). (op. 2016). *The global competitiveness report 2016–2017: Insight report*. Geneva: World Economic Forum.

Stata, R. (1994). Organizational learning – The key to management innovation. In C. E. Schneier (Ed.), *The training and development sourcebook* (pp. 31–42). Amherst: Human Resource Development Press.

Stern, A. D. (2017). Innovation under regulatory uncertainty: Evidence from medical technology. *Journal of Public Economics, 145*, 181–200. https://doi.org/10.1016/j.jpubeco.2016.11.010.

Stilgoe, J., Owen, R., & Macnaghten, P. (2013). Developing a framework for responsible innovation. *Research Policy, 42*, 1568–1580. https://doi.org/10.1016/j.respol.2013.05.008.

The Economist. (2018). *Should the tech giants be liable for content?: Truth and power*. https://www.economist.com/leaders/2018/09/08/should-the-tech-giants-be-liable-for-content. Accessed 1 Mar 2018.

The Guardian. (2017). *Technology company? Publisher? The lines can no longer be blurred*. https://www.theguardian.com/media/2017/apr/02/facebook-google-youtube-inappropriate-advertising-fake-news. Accessed 1 Mar 2019.

van de Poel, I., Asveld, L., Flipse, S., Klaassen, P., Scholten, V., & Yaghmaei, E. (2017). Company strategies for Responsible Research and Innovation (RRI): A conceptual model. *Sustainability, 9*, 2045. https://doi.org/10.3390/su9112045.

van den Hoven, J. (2013). Value sensitive design and responsible innovation. In R. Owen, J. R. Bessant, & M. Heintz (Eds.), *Responsible innovation: Managing the responsible emergence of science and innovation in society* (Vol. 47, pp. 75–83). Chichester/West Sussex: Wiley.

Voegtlin, C., & Scherer, A. G. (2017). Responsible innovation and the innovation of responsibility: Governing sustainable development in a globalized world. *Journal of Business Ethics, 143*, 227–243. https://doi.org/10.1007/s10551-015-2769-z.